RBC

RBC

Tropical Vegetable Production

TROPICAL VEGETABLE PRODUCTION

Raymond A.T. George

Former Senior Lecturer
University of Bath
UK

and

Consultant to the United Nations Food
and Agriculture Organization
Rome
Italy

www.cabi.org

CABI is a trading name of CAB International

CABI Head Office
Nosworthy Way
Wallingford
Oxfordshire OX10 8DE
UK

Tel: +44 (0)1491 832111
Fax: +44 (0)1491 833508
E-mail: cabi@cabi.org
Website: www.cabi.org

CABI North American Office
875 Massachusetts Avenue
7th Floor
Cambridge, MA 02139
USA

Tel: +1 617 395 4056
Fax: +1 617 354 6875
E-mail: cabi-nao@cabi.org

A catalogue record for this book is available from the British Library, London, UK.

Library of Congress Cataloging-in-Publication Data

George, Raymond A. T.
Tropical vegetable production / Raymond A.T. George.
    p. cm.
  Includes bibliographical references and index.
  ISBN 978-1-84593-753-9 (alk. paper)
1. Tropical vegetables–Developing countries. I. C.A.B. International. II. Title.
SB324.56.G46 2011
635.0913–dc22

2010020579

ISBN: 978 1 84593 753 9

Commissioning editor: Sarah Hulbert
Production editor: Shankari Wilford

Typeset by AMA Dataset, Preston, UK.
Printed and bound in the UK by the MPG Books Group.

Contents

Preface

Despite efforts to improve and increase food production and supply by institutional developments during recent decades, there still remains a major food security problem in many parts of the world, especially in the tropics and subtropics.

It is now being increasingly recognized and appreciated that a large proportion of populations in developing countries are reliant on subsistence farm production for their day-to-day food supply and longer-term food security. This food source is frequently referred to as the 'informal sector' of food production. The recognition of this fact was an important outcome of the G8 Conference held in Italy in 2009 and it was also emphasized by President Barack Obama during his address in Accra, Ghana, on 10 July, 2009. The President also stated that 'Aid is not an end in itself. The purpose of foreign assistance must be creating the conditions where it is no longer needed.'

One of the most important aspects of subsistence farming is the day-to-day availability of fresh and stored vegetables and their products. Despite what has been achieved in the past by national and international programmes it is very apparent that we must do significantly more to ensure that the subsistence farmers are able to not only increase their vegetable productivity and food security but that their improved food supply becomes sustainable thereby improving the health, well-being and education of their dependents and overall reduction of poverty. Many of us who have been involved with agronomy and related subjects in tropical environments believe that more effort has to be given to improving food security from the bottom up, that is by improving the information and material inputs such as improved seed, planting materials and technology available to small farmers. There is an urgent need for more farmer training and research which is of direct relevance to improving production on the subsistence farms.

The overall environment is a very topical global subject, coupled with this is the established but still rapidly developing topic of Integrated Pest Management

(IPM). There is a need for farmer training to assist the recognition and adoption of this concept therefore helping the technology transfer of this and other exciting improvements that must be made down to village level.

A subsistence farmer can be defined as one who attempts to produce enough food from a plot of land to sustain the farmer and their dependents throughout the year. By this definition there are no surplus crops or products to market and the farmer is sometimes referred to as a resource poor farmer. Subsistence farmers' main concerns are food security and minimizing risk. They have to rely on one site for continuity of their food supply in contrast to members of the population who are able to purchase from markets.

The gender issue is a very important aspect of subsistence farming. Approximately 50% of food crops in Asia are produced by women, while in Africa women provide about 60% of the agricultural labour force and up to 80% of the total food-producing labour force (FAO, 1998). In addition many women are also responsible for producing the food consumed by their families. It is extremely important from the women farmers' points of view and for their welfare that they become sustainable farmers with appropriate and proportionate opportunities, including access to extension advice and related research findings.

This volume is in two parts. Part 1 comprises six chapters concerning the principles and practice of tropical vegetable production. In Part 2 the crops have been mainly dealt with according to their taxonomy as botanical families, either as single or groups of families per chapter. Examples of the indigenous species which can be regarded as important sources of edible vegetative materials which are not dealt with in the main text have been listed in Appendix 1. Contact details of the main international research stations are provided in Appendix 2. A textbook cannot remain up to date with changes and new developments in crop protection chemicals because pesticides are continually being withdrawn and replaced, in some cases their formulation and approved use is modified, and recommended usage is likely to differ from one country to another.

This book has been written with the hope and purpose that it will be used by technical, college and university students during their studies of horticulture, crop production and agriculture; it is also for students on other allied courses and agriculturists who find themselves needing more vegetable-orientated information in the course of their professional activities. My purpose and objective is to humbly but whole-heartedly assist in the production of extension, advisory and research staff and officers who will be the core of trainers, advisors, researchers and extension workers in tropical and subtropical countries.

I am very grateful to Peter R. Thoday for the very useful discussions and encouragement while I was formulating the scope of the book and for his constructive comments from time to time during its writing.

My thanks to Sarah Hulbert, the Commissioning Editor at CABI, for her patience and ever ready advice while I was preparing the manuscript, also Katherine Dalton, Editorial Assistant, and Shankari Wilford, Production Editor, for so kindly and efficiently dealing with the production side. I am very grateful to Priscilla Sharland who so diligently and efficiently copy-edited the manuscript.

Special thanks are due to our family, Andrew, Christopher, Jane, Joseph and Patrick who helped with the preparation of figures and took care of the occasional computer mishaps which arose during the 11 months of manuscript production and who also, from time to time, improved my computer literacy.

Finally my great appreciation to my wife, Audrey, who over many years has not only tolerated my absences on long-term stays overseas and shorter missions relating to vegetables but has also been very patient while I have been writing this title.

Raymond A.T. George
Bath, UK
May, 2010

Acknowledgements

The author is extremely grateful to the following for permission to reproduce photographs:

Fig. 2.1 Photograph by kind permission of Lucy E. King, the copyright holder.

Fig. 2.3(a) and **(b)** Photographs by kind permission of Alan Spybey, Director of Technology Development, KickStart (www.KickStart.org), the copyright holder.

Figs 5.1, **6.1**, **6.2**, **8.2(a)**, **8.2(b)**, **9.3(a)**, **10.4**, **10.5** and **12.2** Photographs by kind permission of The Real IPM Company (K) Ltd (www.realipm.com), the copyright holders.

1 Introduction

What Is Meant By 'the Tropics' in Terms of Vegetable Production?

If we take a physical and geographical definition of the tropics we find that this area of the world is between the northern limit of the Tropic of Cancer and the southern limit of the Tropic of Capricorn (i.e. between latitudes 25½° N and S of the equator). However, the range of geographical locations suitable for satisfactory growth and production of tropical vegetable species can go both beyond and be limited within this defined area depending on factors such as rainfall, solar radiation and temperature. The environmental limits which restrict the subsistence farmer also interact with the tropical environment.

According to Kelly (1994), the terms 'arid/humid' are defined by reference to rainfall because the records are generally available, although it is recognized that it would be more accurate to take into account evapotranspiration as well. These definitions are:

- Less than 400 mm rainfall/year = arid.
- 400–599 mm rainfall/year = semi-arid.
- 600–1200 mm rainfall/year = sub-humid.
- Over 1200 mm rainfall/year = humid.

These arid/humid definitions are also used as part of the system for providing background information relating to cultivars' performances in cultivar trials and for determining a given environment's suitability for the production of grain legumes (pulses) and seed crops.

The main features which contribute to making a tropical environment are: (i) solar radiation; (ii) temperature; (iii) day length; and (iv) rainfall. The detailed separate and combined effects of these four features have been explained in detail by Cobley and Steele (1976).

The comparison between crop production in temperate regions and the tropics draws attention to some difficulties which the tropical vegetable grower has always had to tolerate. These are especially relevant to subsistence farmers and include:

- The continuously warm soil results in a rapid breakdown of organic matter.
- Tropical rains result in leaching of minerals and nutrients, resulting in some deficiencies or at least low availability of some essential nutrients.
- Inorganic fertilizers are leached very quickly.
- Many harvested leafy vegetable crops deteriorate rapidly as soon as they are harvested and some tropical root crops do not have a very long dormancy period. This jeopardizes their long-term storage potential and use for food security.

The Concept of Subsistence Farming

The earliest style of farming that can be regarded as *subsistence farming* is shifting cultivation. This practice which is now in decline has been regarded in previous centuries as the most widely used technique for claiming and establishing a site which is capable of crop production. The basis of shifting cultivation is that firstly a plot is established by clearing forest or scrubland either by burning, or cutting existing trees and shrubs to within about 1 m of soil level. Some trees or shrubs may be left *in situ* to provide shade or structures for climbing crops; the initial clearing is followed by cultivation of the soil. When the site has proved to be no longer sufficiently fertile for continued cultivation it is abandoned in favour of a new site. Eventually (over a period of years) the site is restored again to scrubland and trees, during which time soil fertility has improved and the cycle is likely to be repeated after a long fallow period. It has generally become accepted in recent years that shifting cultivation cannot lead to the satisfactory development of infrastructures and many national governments either do not allow or at least discourage the practice. Other demands for land arise from the establishment of national parks, game reserves, industrial and urban development, production of crops for food processing and biofuels. The increased populations have resulted in urban development which in turn requires a sustainable food supply.

Definition of a subsistence farmer

Several titles occur in the literature and in day-to-day conversations which loosely refer to subsistence farming. These also include: *smallholder, small farmer* and *woman farmer*. In the context of this volume and its objectives the term *subsistence farmer* includes all of these implied titles. Collectively, including home gardens they embrace the 'informal sector' of the production of vegetables and associated crops. This very significant informal sector has frequently been overlooked, especially regarding the aspects of agricultural and horticultural crop development which have been organized and implemented from the 'top down' on an institutional basis. There is no hard and fast rule as to size of the parcel of cultivatable land that a subsistence farmer cultivates, the size often depends on location, soil or site quality; it may also depend on the number of dependents of the farmer and their concerted activities to cultivate and manage it successfully. The area of a single subsistence farm may range between 1000 and 40,000 m^2 although there is no hard and fast definition of land area.

A subsistence farmer may be defined as one who attempts to produce enough food from a plot of land to sustain the farmer and their dependents throughout the year. By this definition there are no surplus crops or products to market. Thus, also by definition, a subsistence farmer is either not able, or not prepared, to produce a surplus to send or take to market so as to gain any monetary income. In an ideal situation the subsistence farmer would be able to store or process produce for continuity of supply and also to have sufficient food security against vegetable shortages during the high temperature or rainy seasons.

The concept of producing crops for sale, usually via markets, is assumed to imply the development and establishment of a marketing system, although many of the markets may be informal as are 'farm gate' sales or the producer (or member of their family) carrying the surplus for sale at a municipal or other open market. In many of these situations in development we are faced with initially assisting subsistence farmers to being self-sufficient, thus achieving 'food security' for the dependent family unit; this can then be further developed so that a surplus is produced which can be marketed or traded, thereby the subsistence farmer can start on the road to economic growth.

From this concept it becomes clear that while food aid is essential at times of disaster, it is technical aid in the form of inputs such as seed and planting materials, accompanied by technical assistance, which is needed beyond the recovery stage so that a sustainable infrastructure can be developed from the village level upwards. It is this that will assist the grower, his family and his or her community in the longer term.

The constraints facing subsistence farmers are many and can include:

- Minimal land available per farmer, insufficient plot size for sustainability of the family unit. This may arise from inheritance laws, competitive demand for the better crop production sites, over population or insufficient allocation of land compatible with size of the dependant family unit.
- Poor quality site, soil or topography (see 'Optimum Site Requirements' at the start of Chapter 2 of this volume).
- Harsh climatic conditions, resulting in crop failure or poor harvests.
- Frequent pest and or disease outbreaks which are uncontrollable, also mammalian intruders such as elephants or monkeys.
- Individual farmer's physical capability, gender or age.
- Lack of training or suitable technical information to become progressive.
- Lack of satisfactory inputs, such as quality seed, planting materials and crop nutrients (these constraints are especially severe in communities isolated by distance and difficult terrain).
- Subsistence farmers have to rely on one site for continuity of their food supply in contrast to members of the population who are able to purchase from markets, commercial shops or trading establishments which can procure food from different sources, elevations or geographical areas for continuity of supply.

Definition of a subsistence farmer by gender

The physical and other material constraints have been outlined above but the gender issue is a very important aspect of subsistence farming and must always be taken into consideration.

There are many women farmers who by definition are subsistence farmers. The reasons that they are in this situation are numerous, examples include:

- widowhood, or incapacitated husband due to illness or accident;
- single parent;
- husband or partner working away from home; and
- social or religious circumstances.

Approximately 50% of food crops in Asia are produced by women, while in Africa women provide about 60% of the agricultural labour force and up to 80% of the total food-producing labour force (FAO, 1998). In addition many women are responsible for producing the food consumed by their families.

The Importance of Vegetables in the Human Diet

Vegetables and fruit are essential components of the diet and are complementary to the main staple crops which usually form the bulk of the diet. The main staple crop or crops relied upon, or preferred, in a given area include wheat, maize, potato (*Solanum tuberosum*) and rice; sorghum, millets and grain amaranths are also important in some areas. The tropical tubers and/or some of the Andean tubers which are dealt with in this volume as vegetables are also important staples in some environments or communities. The Asian Vegetable Research and Development Center (AVRDC, 1998) produced a plan which included specific programmes such as the inclusion of vegetables in cereal-based systems and year-round vegetable production systems. The plan recognized, and has taken into account, the social and economic values of vegetables.

There are several reasons for growing vegetables but the most important vegetables are essential in the diet as they provide fibre, trace minerals, antioxidants, vitamins, folacin, carbohydrates and protein (Oomen and Grubben, 1978; Gormley, 1989). There is an increased emphasis in developing countries to supplement the staple foods (such as rice, maize and tubers) with locally produced green vegetables such as *Amaranthus* or Chinese cabbage which increase the food's nutritional value. In addition to their nutritional benefits, many of the vegetables used, including 'pot herbs' which may be used in relatively small quantities, make the staples more appetizing and provide extra flavour which is especially important in the hot or dry seasons when vegetables are not plentiful.

The cruciferous vegetables contain glucosinolates which have cancer-protective properties, although the level of glucosinolates vary between different genera and species they each provide some protection (e.g. radish is low in content, but communities consuming significant quantities of radish are thought likely to consume sufficient quantities to provide them with protection). The biochemistry of the vegetable *Alliums* has been reviewed by Brewster (1994) and Block (1985) has discussed their medicinal effects and possible advantages to health.

Edible indigenous plant species

There has been a renewed emphasis on the identification of edible indigenous plant species, especially in areas where the local wild species have a high nutritious value but also fit into year-round cropping systems (Weinberger and Msuya, 2004). This requires identification and growing on the indigenous species for evaluation as illustrated in Fig. 1.1.

Many of the indigenous vegetable species which are not generally considered as major crops have been shown to be both palatable and important for human nutrition, especially when the normally cultivated vegetables are not in season. Examples of some of the relevant indigenous genera and species of importance which are not dealt with in the main text have been listed in Appendix 1 of this volume where the species are listed according their scientific names, botanical family and examples of common names.

Raising the profile of nutrition in research and programme planning

It is important that the nutrition of the consumer be taken into account when planning research and other activities relating to vegetable production. For example, in a breeding programme for improvement of agronomic

Fig. 1.1. Demonstration and evaluation of indigenous vegetables at Arusha, Tanzania.

characters of a vegetable species, the specific nutrient content of the end cultivar can be included in the selection of progeny. Thus, a multi-disciplinary approach to include nutritionists in addition to agronomists and plant pathologists can be very useful to widen any programme's objectives, but it also remains important that the farmers are also included in the evaluation of breeding lines and improved techniques. Enumeration of the different levels of planning and improvement to include nutrition in agronomic development have been outlined by the United Nations Food and Agriculture Organization (FAO, 2001a).

Classification of Vegetables

In order to emphasize the very wide range of vegetables cultivated in the tropics we can list them according to their morphology which is of culinary importance, especially when used in conjunction with a staple food such as rice or maize. This wide diversity is shown in Table 1.1 which outlines the range of botanical species and their morphology regarded as vegetables.

Vegetables can be classified according to altitude, but this is not a clear-cut classification although some authorities attempt to define vegetables as either *lowland* or *highland* species, determined by whether they are cultivated predominantly below or above 1000 m above sea level.

The term *exotic vegetables* is often used in the tropics and subtropics to indicate vegetable species which have been introduced from other parts of the world, and often infers a species originating from temperate regions. It is not a very reliable definition in this context when classifying crop species.

Table 1.1. Examples of the classification of vegetables according to morphology.

Crop	Scientific name	Part(s) with a culinary use
Flowering cabbage	*Brassica chinensis*	Flowering shoot
Lentil	*Lens esculentum*	Seed
Bean	*Phaseolus vulgaris*	Immature seed pod and mature seed
Soybean	*Glycine max*	Seedling
African spinach	*Amaranthus cruentus*	Leaf
Chinese cabbage	*Brassica campestris* ssp. *pekinensis*	Leaf bud
Cucurbita species	*Cucurbita pepo*	Flower, immature fruit
Pumpkin	*Cucurbita mixta*	Mature fruit
Yams	*Dioscorea* species	Root tuber
Potato	*Solanum tuberosum*	Tuber
Onion	*Allium cepa*	Bulb

Future Requirements and Priorities

Many of the current practices adopted by subsistence farmers have a high input of manual labour. There is a need to improve conditions so that technological advances can be adopted by the farming communities. The provision of information and availability of low technology techniques and equipment can help the small farmer make a vital step forward in the improvement of his or her and their dependents' well-being.

There is a greater urgency and need to turn attention to the 'informal sector' of food production in order to reduce poverty of subsistence farmers and their dependents. The need for an increase in 'smallholder' food production in the context of food security is well known. The following points apply particularly to subsistence farmers: (i) crop planning for the season or year ahead; and (ii) continuity of crop supply for available carbohydrate, protein and vitamins.

Systems for continuity of supply for subsistence farmers can include:

- Succession by planned or programmed cropping including successional cropping.
- Use of suitable cultivars for seasons and environments.
- On-farm processing and/or storage of harvested crops, especially for surpluses and gluts.
- Incorporating the best pattern of site use, including intercropping, multi-cropping and rotation.
- Exploring the possibility and reality of developing groups of smallholders and also developing cooperatives at village level.

Secondary uses of crops

Many agricultural crops have a secondary use of direct interest to farmers, for example with rice, the qualities of the straw are also very important to the subsistence farmer in addition to the cultivar's agronomic and culinary qualities. The stubble of cereal crops is ultimately ploughed in or dug in, to incorporate it with the soil, which in time assists the soil's organic status. If the stubble has been grazed then some animal manure has also been added; some straws are better suited for bedding when used in livestock housing than others.

The parallel to this in vegetables is often less obvious. The remains of a vegetable crop following harvest are frequently used as livestock fodder (especially leafy material), directly incorporated in the soil *in situ* or otherwise removed and composted; in this context almost nothing is wasted or lost from the local ecosystem. The remains of maize and Guinea corn crops can be used as a framework for yam vines. The maize or sweetcorn crops grown at a closer spacing for the production of mini-corn can be used as livestock fodder. Break crops, such as a legume or grass cultivated to improve soil fertility, could also be grazed *in situ* thus having a dual purpose of both soil amelioration and stock feed.

Further Reading

Tropical botany

Cobley, L.C. and Steele, W.M. (1976) *An Introduction to the Botany of Tropical Plants*, 2nd
edn. Longman, London.

Nutrition

Asian Vegetable Research and Development
Center (AVRDC) (1998) *Vegetables for
Poverty Alleviation and Healthy Diets*.
AVRDC, Taiwan, 29 pp.
Food and Agriculture Organization (FAO)
(2001) *Incorporating Nutrition Consider-
ations into Agricultural Research Plans and
Programmes*. FAO, Rome.

Gormley, T.R. (1989) Studies on the role of fruit
and vegetables in the diet. *Professional Horticulture* 3, 66–68.
Oomen, H.A.P.C. and Grubben, G.J.H. (1978)
*Tropical Leaf Vegetables in Human Nutri-
tion*. Royal Tropical Institute, Amsterdam
and Orphan Publishing Company, Willemstad, Curacao, 140 pp.

Crops

Weinberger, K. and Msuya, J. (2004) *Indigenous Vegetables in Tanzania*. Asian Vegetable Research and Development Center
(AVRDC) Technical Bulletin No. 31. AVRDC,
Taiwan, 70 pp.

2 Site, Topography, Soils and Water

The subsistence farmer does not usually have the opportunity to choose the site that he or she will be using for vegetable production. However, the main considerations from the point of view of vegetable crop production are discussed below, thus drawing attention to all the prevailing and limiting factors which the farmer has to face. The main requirements considered here include site security, access to markets, topography, soil and water supply.

Site Security

Adequate fencing or other forms of protection are required mainly to exclude stray domestic animals and the smaller wild animals such as feral pigs. However, fencing is unlikely to deter pilferers. More elaborate protection is required to exclude elephants and also climbing animals such as monkeys. Traditionally, a surrounding band of pepper plants has often been used around a vegetable plot to deter elephants. More recently a *beehive fence* using beehives at intervals on the perimeter boundaries has been shown to deter elephants in Africa (King *et al.*, 2009). The study was initiated in Kenya where it was observed that elephants tend to avoid trees which have beehives in them. The beehive fence is illustrated in Fig. 2.1.

The fence is constructed with log beehives hung under small protective thatched roofs. The hive protection 'huts' are placed 8 m apart. An elephant walking between the huts will be less than 4 m from the nearest hive. The beehives swing freely, suspended from tightly secured horizontal fencing wire to the tops of the 2 m poles. The hives are linked to each other with strong, taught fencing wire that hooks to the centre of the permanent wire of each hive and is, crucially, behind the poles on the crop side of the fence. An intruding elephant trying to enter the plot will avoid the complex solid structure of the bee huts and will be channelled between them. As the elephant tries to

Fig. 2.1. Beehive fence to deter elephants. Photograph courtesy of Lucy E. King.

push through the thigh-high wire it causes the attached beehives to swing violently, thereby disturbing and releasing the bees to irritate or sting the elephant. However, if forced, the interlinking wire will break away before the beehive is pulled down; this also prevents elephants being trapped inside the plot as they can break out without damaging the hives. Honey badger attacks on the hives are prevented by nailing a 50–70 cm strip of metal sheeting half way up each vertical wooden post. The farmer has the added advantage of a honey supply as an additional commodity towards food security; this has encouraged and enabled farmers to invest in extensions to the fencing. The system is now being further developed. Kenyan Top Bar Hives which have a queen excluder and are much more productive for getting pure honey from the hives will be used rather than the traditional log hives where the brood has to be almost destroyed to obtain the honey (Lucy E. King, Oxford, 2010, personal communication).

Access To and From Markets

Market access is essential if surplus crops are not going to be processed or stored exclusively for the household's benefit, or for sale or exchanges within the local community. Satisfactory road connections and conditions are important to minimize travelling time and to ensure that the risk of damage and deterioration to crops resulting from poor road surfaces and prolonged heat exposure are minimized. Conversely access to the farmers' villages or sites is important for the supply of farming requisites including items such as fertilizers, crop protection chemicals, seeds, planting materials and replacement parts for machinery.

Topography

It is extremely unusual for land to be flat or 'dead level'. In many cases the slope of the site can be used to advantage for water distribution by channels, furrows or trenches. However, impact from water droplets will not only break down crumb structure and percolate the soil but heavy or continuous rain (or even overhead irrigation) will result in surface panning and also carry soil sediments down the natural slope resulting in serious erosion. Hudson (1957) showed that with a 4.5% slope, 470 t of soil/ha/year can be eroded from a bare soil, but that only 3 t/ha/year was lost from a permanent grass sward. Wrigley (1969) also drew attention to this potential problem relating to water runoff by pointing out that 'If the rate is doubled, it has four times the scouring capacity, thirty-two times the carrying capacity and can carry particles sixty-four times as large'.

Factors which contribute to the risk of runoff include:

- heavy rainfall, its seasonal pattern and occurrence;
- rainfall duration;
- site topography, and this may also interact with the topography of the greater overall location; and
- agronomic practices and land use as applied for vegetable production, including crop stage, plant density and crop coverage of land surface.

Factors which reduce the risk of runoff include:

- The denser-covering vegetable crops, including high plant densities or mixed cropping with more than one crop species.
- Stabilization of the surrounding higher land with trees, plantation crops or grassland.
- Water infiltration rate – low infiltration or percolation rates of water through soils generally lead to anaerobic conditions and result in waterlogging and subsequent root diseases; many vegetables including sweet potatoes, yams and peppers will not tolerate these conditions. This is especially aggravated by impermeable subsoils. A high water table is another factor which can adversely affect drainage in these conditions.

The soil type has a significant effect on water infiltration and drainage:

- Clay soils generally have low infiltration and drainage rates and sometimes drainage is totally impeded.
- Sand and gravel soils have relatively high infiltration and drainage rates, usually resulting in rapid drainage.

Mechanical measures for soil conservation and land formation

The formation of natural land topography can be modified by engineers and/or farmers by a range of systems including terracing, bunding (earth banks), drainage channels across the gradient and surface grading; these are collectively known as *earthworks*. For further details of the construction of earthworks see Webster and Wilson (1980). Other work can include the protection of river banks where appropriate.

The technique of *terrace* construction has been well established in many areas of the tropics, for example terraces have been established in mountainous areas such as Peru and on severe slopes in Malaysia, Indonesia and the Philippines where they provide relatively level plots suitable for crop production although access may not always be easy.

Other measures which can assist the stabilization of sloping sites can include:

- contour ploughing and cultivating;
- inclusion of forage-crop or grazing strips along the contours;
- limitation of grazing on severe slopes; and
- belts of small trees and/or shrubs planted along the contours.

Soils

A smallholder or subsistence farmer in the tropics is extremely unlikely to be able to choose the soil or site on which to grow vegetables. There are four main types of tropical soils (usually referred to as *soil orders*), which classify soils according to their origins, these are: aridsols, oxisols, entisols and alfisols.

Aridsols are soils which have developed in a very dry environment. They generally have a very low humus content which accounts for their light colour. Many areas where agriculture is dominated by nomadic stock grazing have this soil order. The low humus level results in lack of nitrogen therefore crops growing on these soils will require available nitrogen during their growth and development although there are usually sufficient levels phosphorus, calcium and potassium.

The *oxisol* soil order has developed in tropical and subtropical latitudes where there are both high temperatures and rainfall such as central South America, Equatorial Africa and parts of South-east Asia. Although there can be organic matter present, most of the nutrients have been lost by leaching. This soil order does not have any obvious profile or horizons. The rate of weathering associated with high rainfall has reduced the nutrient availability and they are not really suitable for vegetable crop production although some subsistence farmers are obliged to cultivate them.

Entisols are relatively 'young' or immature soils and do not display clear horizon differentiation. This soil order has usually been formed as a result of windborne sediments, water or ice deposition especially along the banks of rivers or below steep slopes where the soil results from erosion. The ultimate characters of soils formed in this way depend on the character or quality of the materials from which they are derived.

The *alfisols* have been formed from considerable weathering of forest vegetation and associated materials. These soils contain significant quantities of aluminium (Al) and iron (Fe) hence the name alfisol. They have clearly formed horizons of which the surface one usually has a light colour. They are relatively high in base cations. This soil order usually requires additional liming materials in addition to adequate phosphorus and potassium for *effective* vegetable crop production. Any shortfall in organic materials for sufficient humus content for vegetable production will need to be corrected by frequent addition of suitable materials.

Soil texture

In addition to the rather broad classification of the four soil types outlined above which classifies a soil according to its origins, any given soil can also be classified according to its physical properties. When a moist soil sample is rubbed between finger and thumb, an assessment can be made of the presence of sand, silt, clay and organic matter. With most samples only the first three of these can be felt as the humus content is low in proportion to sand, silt and clay.

There are three constituents, or mineral fractions, whose proportions by weight tell us a given soil's physical characters. The three constituents are: sand, silt and clay. Determining their relative proportions gives us the texture of an individual soil and provides us with an insight into the soil's structure and chemical properties. These properties are further modified or influenced by cropping, cultivation methods and organic matter. Each of these three constituents imparts the following characters to soil texture:

- *Sand* imparts grittiness which can be felt when handling a soil sample. The actual particle size will vary according to the sand's origin. Generally the higher the proportion of sand in a soil the lower the cohesion; this may be regarded as the force that resists the breaking down of the soil into smaller particles by tilling operations.
- *Silt* imparts a silky or soapy feeling, but does have some cohesive properties.
- Clay, when moist, imparts a polished and sticky feel.

The other, non-mineral, constituent is *organic matter*. This originates from plant and animal residues and is therefore variable in size, origin and degree of decomposition. It generally feels soft and imparts cohesive properties to a soil.

Different vegetable crop species tend to have a preferred soil type which is best suited to their optimum crop production. Appropriate or optimum soil types for individual crops are given in Part 2 of this volume, although it must be appreciated that the subsistence farmer rarely has a choice. From the 'field factor' point of view, soils can be regarded as either mineral or organic.

- *Mineral* soils vary in texture and may be described as *sandy* (at least 85% sand), *loam*, *silt* or *clay* (at least 40% clay).
- *Organic* soils may be described as *well decomposed* or *not well decomposed*.

General implications of soil structure for vegetable crop production

A natural soil consists of clay, sand, silt and organic matter; these constituent particles do not exist as separate entities but as less stable soil crumbs or aggregates. The size, distribution, shape and stability of these secondary aggregates is referred to as *soil structure*.

The importance of the clay fraction derives from its large contribution to the total surface area of soil particles (Flegmann and George, 1975). The very dilute soil solution requires continuous replenishment with nutrients from the soil solids. The retention of nutrient ions by the surfaces of clay colloids mainly depends upon the presence and distribution of electric charges on their surfaces. The clay fraction provides a large surface area for the reserve of nutrients for plant growth. A soil which has a low proportion of clay therefore has a low potential fertility, and any fertilizers applied will be liable to leaching. However, plant roots require oxygen (obtained from the air), water and physical space for penetration as well as anchorage; thus soils with a high proportion of clay may be considered unsuitable for vegetable production. Winter (1974) reported that although clay soils are able to hold more water at field capacity (the maximum amount of water that a particular soil can hold without drainage) than any other texture soil, the water in the small pores is held at higher tensions than in the pores of silty and sandy soils, thus a lower proportion of water is available for plants to extract.

Acidity and alkalinity of soils

The natural soil acidity can range from a pH of 3.5 to 9.5; a pH below 6.0 is defined as acid, pH 6.0–7.5 as neutral and a pH above 7.5 as alkaline. The majority of plant species cultivated as vegetables thrive best in soils with a pH between 5.5 and 7.0. Soil pH has a significant effect on the availability of macro- and micronutrients essential for satisfactory plant development and growth.

The soil pH can be modified by the application of liming materials to increase alkalinity; or by application of elemental sulfur to increase acidity. The use of certain inorganic fertilizers over several seasons can also affect soil pH, for example applications of ammonium sulfate, often applied as a nitrogenous fertilizer, would be expected to increase soil acidity.

Salt levels and management of subsoil salt levels

Some soils contain levels of salts which are sufficient to regard the soil as saline. The high soil salt levels tend to occur in arid and semi-arid areas where soils are often naturally saline; this is the result of evaporation exceeding precipitation. In this case the salts are derived from lower soil horizons. Other causes of soil salinity may be the application of saline irrigation water or from residues of fertilizer application. In the case of the subsistence farmer it is more likely to be a result of soils being naturally saline or due to saline irrigation water rather than from fertilizer residues, as the latter tends to be associated with soils in which there has been intensive mono-cropping with high fertilizer levels having been applied in intensive crop production regimes. The other possibility of soil salinity is where there has been seawater flooding.

A saline soil is defined as a soil with a specific conductivity of 4 mho/cm corresponding to approximately 3000 parts per million of salts. Soils with conductivity levels lower than 4 mho/cm are considered to be normal, whereas those with higher levels are regarded as salty. Table 2.1 lists vegetable crop species according to their tolerance to soil salt level. A detailed account of salt tolerance in plants has been provided by Berstein (1970).

Methods of improving nutrient status and soil texture

Livestock and their relationship with the small farm
All farmers who cultivate arable crops, including vegetables, are concerned with maintaining or even increasing the soil's humus content. In addition to other methods of increasing organic matter content there is the possibility of regularly incorporating animal manure obtained from overnight or winter housing of livestock (e.g. goats, sheep, water buffalo or cattle) resulting in the availability of *farmyard manure* (FYM). This is incorporated during site preparation or manure deposited *in situ* from animals grazing on leys or short-term forage crops grown on the site. Figure 2.2 illustrates the addition and incorporation of organic materials during site preparations.

Manure derived from small animals
Some farmers keep ducks or chickens but unless the birds' housing is such that their droppings, together with litter material such as straw, can be collected in bulk the available quantities appear to be negligible. However, if the birds are penned on arable parts of the farm their contribution to soil fertility will be significant over time. Poultry manure varies between samples; generally its organic content is approximately 20% nitrogen, present as undigested food, and the remainder is of urinary origin (approximately 60% uric acid and 10% ammonium compounds). If the manure is stored it heats up resulting in volatilization and subsequent loss of the ammonium salts. It is therefore generally recommended that the fresh manure is spread on to a fallow soil surface and mixed in but not sown or planted for approximately 2–3 weeks

Table 2.1. Examples of vegetables' soil salt tolerance.

Salt level	Water salinity (mho/cm) at 25°C at which yield decreased by 10%	Vegetable crop that can tolerate this level
High	8	Red beet
Mid	3	Brassicas, cucurbits, Irish potato
Moderate	2.5	Sweetcorn, sweet potato
Low	2	Pepper, radish, onion
Very low	1.5	Legumes (beans)

Fig. 2.2. Subsistence farmer and son incorporating bulky organic manures during site preparations.

to avoid ammonia scorch of plants or seedlings; during this time rain or irrigation will wash the nitrogen and ammonium compounds into the soil.

Wilson and Brigstocke (1981) outline the relationship between crops and animals in small-scale vegetable production systems. They cite rabbits, guinea-pigs and capybara (a South American rodent of approximately 1 m length) as examples of small lean-meat-producing animal species which have high reproduction rates, low capital outlay, are efficient converters of fibrous vegetable waste and have the ability to cope with the seasonality of vegetable production due to their short breeding seasons. However, social and religious factors must be taken into account prior to making dogmatic recommendations for livestock production in any community; an interesting review of the use of a wide range of rodents in the human diet has been made by Fielder (1990), who discusses traditions and taboos relating to rodents as possible sources of food.

Alternative organic manures
There is a very wide range of organic manures, their availability varies from place to place and they include bird or fish guano, composted town waste, wool shoddy and organic materials produced from animal waste. The production of compost 'on farm' is widely practised. The operation involves the routine collection of waste plant and organic household materials and collecting them in a 'compost heap'. The heap should be sited downwind from the dwelling, in a shady position.

There are two main types of compost heap, the choice of which depends on the amounts of rainfall: (i) the pile method which is generally adopted in areas of high rainfall; and (ii) the pit method, which is used in areas of low rainfall.

Whichever method is adopted, the principle of composting is generally the same. The compost heaps are built up in alternate layers of different types of materials. It is usual to start the 'heap' with a layer of coarse materials such as maize stems or light branches and shoots of a compostable leafy tree. The next layer is green material such as weeds, grass or plant trimmings, next follows a layer of animal manure, this is usually followed by a layer of fire ashes and then by a thin layer of topsoil or compost from a previous well-composted heap. This layer sequence can be repeated, and when more compostable material becomes available it is added to the heap; otherwise a heap is constructed in a single operation. Some farmers turn the heap developed by the pile method over after a fortnight or so. It is important to continue adding water when conditions are dry. The addition of urine from time to time will also be beneficial. Depending on the local conditions and exact proportions of the various constituents a heap constructed in a single operation can be ready for use in approximately a further 2–3 weeks following turning. The green materials provide the nitrogen while the wood ash provides some potassium and also charcoal which neutralizes any acidity that arises from the animal manures. The added soil and remains from an earlier heap provide essential microorganisms. The final height of a pile should not exceed approximately 1.5 m and the ground area covered by a rectangular heap should not exceed an area equivalent to 1.5 × 1.5 m. The top of the heap should be kept covered with an old plastic sheet or large leaves from plants such as plantain or banana.

Mulching

The materials used and mulching methods adopted in vegetable cultivation vary throughout the tropics. A mulching material can be defined as 'any material which is applied to the soil surface' and may be:

- *Manufactured materials* – including black polythene, shredded paper and metal foil. However, these are either difficult to obtain or too expensive for many subsistence farmers. A mulch of black polythene for tomato production is illustrated in Fig. 10.2.
- *Natural materials* – including straw and plant debris from a range of crops, usually of local origin (e.g. rice straw, chopped maize-plant remains), hoed off weeds (ideally hoed before they set seed), plant trash, composted plant and/or animal manure, or tree leaves (e.g. bananas and plantains). Whatever materials are used, consideration should be given to avoid transfer of pests and pathogens from related crops. The use of some dried mulching materials, such as straw, may cause a short-term nitrogen deficiency, and this should be taken into account when deciding on fertilizer regimes.

Advantages of mulching include:

- Weed control by suppressing weed development – this is important throughout the duration of the crop, but is especially important for the reduction of interspecific plant competition.

- Shading germinating seeds or young seedlings (seeds of some crops, such as lettuce, can be thermo-dormant, which can result in sporadic or late germination) – a thin mulch layer of a material such as rice straw can assist the reduction of soil temperature and excessive water loss following seed sowing.
- Modification of soil temperature – this can be a method of reducing the risk of, or even controlling, soilborne wilt pathogens (e.g. Fusarium wilt of tomatoes) which occur at higher soil temperatures. The mulch provides a cooler root zone, especially near the surface where a new adventitious root system can develop.
- Water conservation by reducing water evaporation from the bare soil, preventing soil capping and reducing water runoff.
- Reduction of the soluble-salt levels that result from salt deposits in the upper soil layer as a result of excessive evaporation.

The longer-term effects of mulching include increasing the soil's level of organic materials and a reduction in leaching of soil nutrients. The soil organisms are considered to be more active under mulches, partly as a result of increased soil moisture and partly resulting from the increase in organic materials; however, the organisms responsible for nitrification are less active in areas of high evaporation. Adding to the evidence that mulching with organic materials has a valuable long-term advantage of reducing nutrient leaching in addition to its immediate and shorter-term effects is the work of Gillman and Bell (1972). They showed that the properties of weathered organic soil materials should be both conserved and increased because their negative charge is essential for cation retention of nutrients.

Water Supply

Satisfactory supply of water for irrigation

The availability of irrigation water is an essential requirement for vegetable production, especially in arid and semi-arid areas. In the choice or evaluation of a site all possible constraints of water availability should be taken into account, including restrictions on amounts or times that water may be extracted which may be imposed. Supplies of irrigation water may be obtained from the following sources, subject to these possible restrictions:

- Rivers and streams – possible limitations may result from seasonal flow, long-term drought or restrictions imposed by cost and/or legal rights to draw irrigation water.
- Groundwater – the water is pumped up from boreholes or wells. This has become important in arid areas. However, it must be borne in mind that the extracted water has often collected over a long time and if the groundwater is not replenished by percolation from precipitation, schemes based on this source will not be sustainable in the longer term.

- Collection from precipitation and/or melt from winter snow in mountainous areas – opportunities for this arise with the construction of dams in valleys. These are usually regional schemes and unless done with the specific advantage of the small farmer in mind may not always be of direct benefit to the subsistence farmer.
- Diversion or blocking of rivers to form storage areas with irrigation available for small farmers – examples include the Gezira and Rahad Schemes in the Sudan. This type of scheme is usually managed by a government agency or corporation. Some of the drawbacks experienced by the Rahad Scheme have been outlined and discussed by Sargent *et al.* (1987). Smaller-scale schemes constructed and organized by farmers on a very local scale can create smaller storage volumes for local or individual farmer use.

Quality of irrigation water

The quality of available irrigation water is, in many ways, as important as quantity. The following aspects should be taken into consideration:

- Chemical content – this will affect the soluble salt level of the water and could render the water source too saline for irrigation purposes.
- Toxic materials in solution or suspension – there are several toxic materials, such as boron compounds, which can be at toxic levels in water supplies. Water obtained downstream from industrial processes may contain a range of toxic chemicals.
- Materials in suspension – these include sediments formed from a range of items, including algae and silts which will block sprinkler and spray line distribution systems, also drip and trickle systems resulting in a very uneven water distribution pattern on the irrigated site.
- Pathogenic organisms which affect humans – many of these are distributed by water, for example *Schistosoma* species responsible for bilharzia (OAU, 1970) and *Onchocera volvulus* which is a parasitic nematode causing river blindness (WHO, 1973).
- Some of the human gut-infecting bacteria can be present in irrigation water – these are readily transmitted to the consumer via drinking water, uncooked salad crops or partially cooked vegetables.

Irrigation and methods of application

It should be appreciated that irrigation does not counteract any other shortfalls in good vegetable crop husbandry; for example a crop may be wilting as a result of nematodes or wilt diseases and these conditions will not be corrected by irrigation. Vegetable crops are generally very sensitive to water deficiency compared with other crop groups, and this is particularly important for those crops which produce a fruit (e.g. tomato, cucurbits and leguminous crops) as they have to pass through the moisture-sensitive stage of anthesis in order to produce the required vegetable.

There are a very wide range of methods and systems for the distribution and application of irrigation water. However, only those which may be suitable for small-scale vegetable production are discussed in this volume. There is a saying among irrigation technicians and agronomists that 'plants respond to water, not to the method of application'.

Surface application

There are several methods for surface application, but the one most widely used is open channels or *ridge and furrow*. The ridge and furrow system is widely adopted for crop species with relatively large plants such as tomatoes, eggplant and cucurbits. The site should initially have been graded in order to obtain an even flow and distribution. Prior to planting out or seed sowing, water is run into the channels and an optimum top waterline determined. This brings the soil up to field capacity and also ensures that each plant will be at the same level in relation to the irrigation level. Transplants or seeds are placed just above the waterline. In some situations the seed is sown or transplanted on the top of the ridge. The water is supplied directly to the furrows via siphon pipes or enters the furrows via a smaller channel parallel to the main header. In this latter situation, earthworks adjusted by workers with spades control the flow to groups of furrows. Irrigation via furrows causes less 'splash' on to crops although the total volume of water applied per unit area of cropped land is usually greater than some other distribution systems. The fall of the channels is important: minimum erosion occurs with a fall of 0.1%, a fall of 0.3% is the maximum before erosion becomes serious while 0.5% is the maximum fall provided the volume of water flowing is kept to a minimum.

Trickle or *drip systems* are very economical with the total volume of water used, and can also be used for the application of liquid feed or some pesticides when used with a 'diluter' which works on a venturi system. These systems generally assist the retention of crop protection chemicals applied to foliage and are very useful in areas where the soil salts are rising from the subsoil (provided that the irrigation water has a lower and tolerable salt level). From the subsistence farmer's point of view these systems can have the following drawbacks:

- High capital outlay.
- Tailor-made set-up which usually does not allow flexibility for different spacings when subsequently used for a crop with a different plant density.
- Requires an efficiently filtered water supply and regular 'end-of-crop' cleaning and flushing to remove silt, sand, algae and precipitated chemical deposits which remain in the pipelines and emitters.
- Plastic and synthetic tubes can be subject to gnawing and other physical damage by rodents.

There is a very wide range of methods of *overhead* or *sprinkler systems* of irrigation. They can either be implemented by hand labour or mechanized. The most advanced or large-scale techniques are not discussed here because either their investment cost or scale of operation prohibits their adoption by subsistence farmers. Advantages of sprinkler systems are:

- The shallow fall in land gradation for successful furrow systems is not so critical.
- They generally use less water than ridge and furrow irrigation per unit area.
- They are labour saving, especially on larger areas.
- They remove windborne deposits from edible crops produced in dusty environments thereby increasing photosynthesis and also market presentation of edible crops.

The disadvantage of sprinkler systems is they can be responsible for the spread of foliage pathogens in the humid tropics, especially in crops of eggplant, tomatoes and peppers.

There are many traditional and less sophisticated methods of irrigation some of which are still in use where farmers are unable to afford adoption of more recent developments or techniques. These traditional methods include:

- Overhead watering from water cans – this method requires access to wells or storage tanks from which the cans are refilled but is generally considered to be labour and water efficient for low density plantings. It is still widely used for seedbeds or young plant establishment after planting out.
- Scooping water from canals running in parallel to raised beds – this system is very popular in areas with a high water table such as parts of Thailand where the canals are alongside the plant beds. Hand-operated light wooden scoops or hand-guided floating pumps are used to distribute the irrigation water onto the beds.
- Hosepipe with rose – this requires a pressurized water supply and is very useful for nursery beds. Although subsistence farmers are unlikely to have pressurized mains-water supplies available on site, an excellent supply can be obtained from a well from which the water is lifted by a treadle pump as illustrated in Fig. 2.3.
- Basin irrigation – in this method the vegetable crops are grown in beds which are surrounded by a low wall of soil. The irrigation water enters the bed from a canal at the high end via a temporary cutting in the soil bank or via siphons; each individual bed is usually allowed to be flooded before the water supply is either diverted to another basin or stopped.

Technical low-cost developments in irrigation equipment which assist subsistence farmers include simple irrigation kits such as *trickle irrigation sets* and also *sprinkler systems* that are available for incremental development on smallholdings. The area which each package can usually deal with is usually up to $25\,m^2$. The kits can be assembled on site and subsequently added to in increments over time as the farmer progresses.

The development of the innovative *treadle pump* is an excellent concept of a low-cost piece of equipment which illustrates the value of design simplicity and application of intermediate technology. The MoneyMaker treadle pump shown in Fig. 2.3 is produced and marketed by KickStart. This piece of

Fig. 2.3. (a) The MoneyMaker treadle pump being operated by smallholder farmers in Kenya. (b) A close-up view of the MoneyMaker pump. Photographs courtesy of Simon Mugo, Technical Designer, KickStart (www.KickStart.org).

equipment has been developed as a human-powered water pump using the principle of lifting water from below ground level by twin pistons and cylinders. The application of the operator's legs with their powerful muscles, via a treadle actuated by the operator's feet, provides a significantly greater energy input than hand pumping. A very important feature of these pumps is their low cost and they are therefore generally within financial reach of subsistence farmers. As it is necessary that the piston seals function correctly, there is water over the top of the piston disk and seals. A simple one-way valve is incorporated in the piston disk which allows water to leak through the disk on the pressure stroke (but to seal the suction stroke). Consequently, there is always a small amount of over flow in the pump's operation. In order for the cost to remain low, the leak valve consists of a simple hole and rubber seal, and for there to be enough sealing water when the pump has a low outlet head (for example with a short hose, irrigating nearby plants at the same elevation) the hole has to be large enough to provide sealing water at low pressure. Consequently, when the pump is at its maximum stretch (i.e. 7 m vertically or over 100 m horizontally, or a combination of the two) the high pressure generated causes more water than necessary to flow back through the leak valve, as seen in Fig. 2.3(b). However, the water delivered to the field remains more than adequate (up to approximately 0.8 ha of land can be irrigated), and any excess water from the pump is easily returned directly to the well source without loss (Alan Spybey, Nairobi, 2010, personal communication).

Many of the newer advances suitable for adoption by subsistence farmers have been very successful; their success has resulted from being low cost, simple design and, once a local distributor is established, spare parts and accessories should be available locally on a reliable basis. A comprehensive account of smallholder irrigation technology including low-cost technologies has been produced by FAO (2001b).

The pitcher irrigation system
This is a very ancient system which is thought to have originated in Persia where water supplies from mountain sources were often available from hand-dug underground channels known locally as kanats. Kanat water-supply systems were developed along the Silk Route and evidence of these ancient systems can be found in Iran, the Middle East, Pakistan and China. In addition to providing domestic water sources, water would be lifted in containers and distributed in the area to be irrigated via pitchers which had been buried vertically, up to the pitcher's neck. The pitcher system, which is still used in some Middle Eastern areas and desert areas of India and Pakistan, uses unglazed earthenware pitchers. Rainwater or well water is generally used at the present time. The performance of different size pitchers has been investigated by Siyal *et al.* (2009). The authors found that a small pitcher, half the size of a larger one, but with double the hydraulic conductivity, produced approximately the same wetting front as the larger pitcher.

Water-sensitive stages of vegetables

The majority of vegetable species are annuals or biennials and are generally shallow rooted compared with many perennial crops. Root activity can be encouraged by deeper cultivations and also ensuring that there is an adequate supply of plant nutrients. A review by Salter and Goode (1967) found that most annual crops have moisture sensitive stages from flower initiation, during anthesis, and in some species it continues through to fruit and seed development. In the case of legume crops, provided there is sufficient water available prior to flowering for plant growth to proceed without permanent wilting, there is little influence on crop yield. However, this crop group is very sensitive to moisture stress during anthesis when adequate soil moisture, supplemented by irrigation, will generally provide higher yields. This effect in legumes is due to a reduction in root activity when the crop is in flower. The fresh market yields of the leafy vegetables is generally in proportion to the amount of water received to match their requirements, although the most critical stages can be during crop establishment (especially following transplanting) and during the final period coming up to their harvest.

Further Reading

Fielder, L.A. (1990) Rodents as a food source. In: Davis, L.R. and Marsh, R.E. (eds) *Proceedings of the Fourteenth Vertebrate Pest Conference*. University of California, Davis, USA, pp.149–155.

Food and Agriculture Organization (FAO) (2001) *Smallholder Irrigation Technology: Prospects for Sub-Saharan Africa*. FAO, Rome, Italy.

Kay, M. (2001) *Smallholder Irrigation Technology Prospects for Sub-Saharan Africa*. International Programme for Technology and Research in Irrigation and Drainage (IPTRID). Food and Agriculture Organization (FAO)/IPTRID, Rome. [This is an additional edition of the text above FAO (2001).]

Kay, M. and Brabben, T.E. (2001) *Treadle Pumps for Irrigation in Africa*. Food and Agriculture Organization (FAO), Rome, Italy.

Roberts, E.H., Summerfield, R.J., Ellis, R.H., Craufurd, P.Q. and Wheeler, T.R. (1997) The induction of flowering. In: Wien, H.C. (ed.) *The Physiology of Vegetable Crops*. CAB International, Wallingford, Oxon, UK, pp. 69–99.

Salter, P.J. and Goode, J.E. (1967) *Crop Responses to Water at Different Stages of Growth*. Commonwealth Agricultural Bureaux, Farnham Royal, UK.

Shively, G.E. and Coxhead, I. (2005) *Land Use Changes in Tropical Watersheds: Evidence, Causes and Remedies*. CAB International, Wallingford, Oxon, UK.

Taylor, A.G. (1997) Seed storage, germination and quality. In: Wien H.C. (ed.) *The Physiology of Vegetable Crops*. CAB International, Wallingford, Oxon, UK, pp. 1–36.

3 Site Management, Seeds and Types of Cultivars

Environment Modifications to Enhance Crop Production on Individual Farm Sites

Between and within each of the subsistence production areas local gradations of climate and microclimate can be expected and experienced, these variations result from differences in altitude, topography, proximity to the coast and the presence of natural shelter belts such as woodland or plantation crops. The microclimate in these areas may be further modified and enhanced in one of several ways. These include the use of windbreaks and temporary shelter.

Effects of windbreaks and temporary shelter

The prime purpose of a windbreak or shelter belt is to reduce the wind speed rather than to stop it. Windbreaks or other forms of shelter are especially useful for vegetable crop production where individual blocks or plots are normally significantly smaller in land area than those of many agricultural crops. The advantages include:

- Water loss by plant transpiration and evaporation from the soil is reduced.
- Uneven distribution from overhead irrigation systems is reduced.
- A reduction of drift when spraying pesticides.
- There is less leaf damage by bruising and the protective waxy layer on the leaf surfaces of genera such as *Brassica* and *Allium* remain more intact.
- Windbreaks can reduce the incidence of scorch from windborne salt which originates from sea spray in coastal areas.
- Reduction of the deleterious effects resulting from hot dry winds on some crops, plant-raising areas and exposed nurseries.

- Reduction of damage to temporary structures erected to provide shelter or shading for seedling and plant propagation areas.
- Soil erosion from 'blowing' is reduced which also reduces the risks of wind-borne soil pests and pathogens.
- Deposits of wind-laden dust or sand particles on fruit and leaves are reduced.
- An improved microclimate will enhance flower development, increase pollinating insect activities and therefore pollen transfer in entomophilous crops such as *Cucurbita* species. The reduction of physical damage to flowers will also result in more fertilization and improved fruit set.
- Improved microclimate for predatory insects, this is an important component of Integrated Pest Management (IPM).

The principle of windbreaks

Solid barriers such as walls divert the wind flow but cause turbulence which can damage the plants. However, windbreaks that offer approximately 50% obstruction provide a relatively extensive shelter with the minimum of wind gusts. A windbreak can decrease the wind speed in the horizontal direction downwind for a distance equivalent to up to 30 times its height although most shelter is within a horizontal distance on the leeward side of approximately ten times the height of the shelter.

Types of windbreaks

There are several different types of windbreaks but in principle they are either *permanent* or *temporary*.

Permanent windbreaks

These are provided by planting lines or belts of single or mixed species of trees which are tolerant of local conditions and have a growth rate suited to quick establishment. A wide range of species is used: evergreens such as conifers and *Eucalyptus* spp. and species such as *Populus* spp., *Salix* spp. and *Tamarix* spp. Tolerance of the species to salt should be taken into account where high soil salinity is known to be a potential problem. In many areas appropriate species are planted either along water courses and irrigation channels. Larger land areas can been subdivided into appropriate size plots by lines of trees.

Possible disadvantages of living windbreaks are:

- Roots of the species used may enter water courses and drains with the result of water loss, blockages or structural damage to pipes and conduits.
- Dense windbreaks adjacent to the vegetable production area are likely to reduce the available photosynthetic light and will compete with the vegetable crops for water and nutrients.
- Living windbreaks can be host to a wide range of pests and pathogens. For example some *Populus* spp. are alternative hosts to the lettuce root aphid (*Pemphigus bursarius* L.), also *Salix* spp. to the willow-carrot aphid (*Cavariella aegopodii* Scop.) which is the vector of carrot motley dwarf virus.

- The reduction in wind speed may allow pests such as red spider mite, aphids and whitefly to colonize and settle.
- Dense rows of shrubs and trees can provide a suitable habitat for undesirable bird species which are likely to feed on the crops.

Temporary windbreaks

These are often referred to as short-term windbreaks and are either living materials or manufactured materials. Living plant materials are usually another annual crop grown for its own value and with the dual role of providing shelter for an adjacent crop or crops. Examples depend on the farming system but *Zea mays* L. (either maize or sweetcorn), *Helianthus annuus* L. (sunflower), cultivated *Sorghum* spp. or narrow strips of other cereal crops are often used. Occasionally climbing vegetables, such as *Phaseolus* species or cucurbits, grown on a framework can be positioned so as to provide shelter for other more wind-vulnerable crops.

Other forms of temporary shelter can be provided by erecting screens of manufactured materials such as plastic mesh or hessian. Plastic materials should contain ultraviolet (UV) light inhibitors to prolong their useful life by minimizing UV degradation. Some types of locally available natural materials can be very suitable when loosely woven. These types of structure are relatively low in height, usually no greater than 2 m, but have the added advantages of immediate effect, do not compete with crops for water and nutrients or normally support pests and pathogens. The optimum openness of a screen's porosity to wind movement is 50% material and 50% spaces, often referred to as a '50:50' screen.

Cut and carry shrubs and trees

Cut and carry is the term generally given to trees or shrubs grown for a supply of animal fodder and/or fuel wood. In some respects they can also provide shelter and protection to vegetable crops depending on where they are situated. They should be sited so as to reduce the risks of shading annual vegetable species and root competition. Ideally, the species used should be suited to the local climate and have a dual or even a multiple role by providing fuel wood, animal fodder, soil nitrification (by leguminous tree species) and composting materials.

Examples of species used for 'cut and carry' in different geographical areas are:

- *Boehmeria nivea* – China grass, a shrub used in South-east Asia;
- *Leucaena glauca* – lead tree, originating from tropical America;
- *Tamarindus indica* – tamarind, used in tropical Africa; and
- *Gliricidia sepium* – used in tropical America and the West Indies.

The Role of Seed in Agricultural and Horticultural Development

It is frequently said that 'Seed security means food security'. Many vegetable crops are grown from seed. In this context, 'seed' can also include plantlets or propagules produced by vegetative propagation. It is important that growers, economists, planners, administrators and politicians who are responsible for agriculture and horticulture appreciate that seed is the starting point for the production of many cereal, arable and vegetable crops. The better all aspects of seed supply and seed quality are, including cultivar suitability, the better the chance the farmer has of a successful crop.

The seed industry and seed supply for smallholder farmers

Farmers can ensure a supply of seed for the production of their crops in the following ways:

- Saving seed from their own crop or crops; usually referred to as 'on-farm seed production'. The exchange of 'on-farm' seed between neighbouring farmers can be included in this category.
- Obtaining some, or all, seed requirements from national or regional government seed production centres.
- Buying in seed, hopefully from reliable seed merchants or authentic distributors.
- Receiving seed as a donation from international aid programmes and non-governmental organizations (NGOs), usually as a relief operation resulting from natural or manmade disasters. It is important that 'gifted', 'donated' or subsidized seed does not jeopardize the development of a commercial seed industry which should become the mainstay of a sustainable seed supply. Once a stable seed supply has become established which is capable of providing farmers with the seed quality and cultivars that they require, national government resources can be concentrated on the various aspects of seed quality control.

There are no available figures stating the percentage of vegetable crops in the world grown from farmers' own saved seed, but some estimates suggest that in some developing countries as much as 90% of vegetable seed requirements are saved from farmers' own crops. It is very common for farmers to save their own seeds of the supplementary staple crops, such as cowpea, millet and indigenous vegetables; especially seeds of the legumes which are bulky and also have a low seed multiplication rate per generation. The proportion of 'on-farm seed' not only depends on the level of seed industry development in a country, but for some vegetable crops it also depends on how easily sufficient quantities of seed can be produced in the local environment, and also how isolated by distance and/or topography the growers are. The amount of extra work a grower is prepared to undertake to produce his or her own seed is also a factor, but their economic situation may dictate it.

All vegetable enterprises depend on seed which is at least of satisfactory quality. There are several factors involved in the determination of 'seed quality' and these are usually referred to as *seed quality attributes* which are: germination and vigour, genetic purity (often referred to as 'trueness to type'), mechanical purity, seed health and seed moisture content. Each of these attributes can be influenced and/or affected by the original mother seed, and a range of factors that influence the development of the produced seed on the mother plant. These factors are referred to as *determinants* as outlined in Table 3.1.

Table 3.1. Seed quality attributes and their determinants.

Seed quality attributes	Pre- and postharvest determinants and/or operations influencing seed quality attributes
Germination and vigour	Seed ripening conditions on the mother plant Postharvest ripening Drying Processing (avoiding mechanical damage) Storage, including packaging and container environment Adverse and/or untimely seed treatments
Genetic purity (trueness to type)	Accurate labelling at all stages and times Purity of original stock and/or basic seed Control of ground keepers (volunteer crops) Isolation (in time and distance) Roguing efficiency, at appropriate morphological stages Implementation of cultivar descriptions
Mechanical purity	Processed seed free from seeds of other crops, free from other impurities, weeds or parasitic plant species Efficient winnowing, seed cleaning and correct use of appropriate specialist seed cleaning machines, depending on contaminants present in the seed lot
Seed health	Stock seed free of seedborne pests and pathogens Adequate control of volunteer plants in the field which are alternative hosts to pests and pathogens Roguing out infected mother plants as soon as identified Timely and efficient seed treatments
Moisture content	Harvesting conditions Timely harvesting and drying

Quality Control of Seed and Planting Materials

Seed

When a vegetable grower has decided on the most appropriate cultivar to use for a specific purpose it is desirable that he or she be supplied with genuine seed of that cultivar and that the seed lot has the best possible attributes as listed in Table 3.1. A range of schemes which certify the authenticity of the seed marketed have evolved in different parts of the world, each provides some form or level of assurance that the seed supplied is actually of the cultivar that it is claimed to be. In theory this ensures verification of the seed stock without the farmer or grower having to see the resulting crop derived from it. However, in practice, the quality assurance depends on both the rigour of the system as well as the discipline and efficiency of the different parties responsible for implementing it. Kelly (1994) has clearly described and discussed policies for seed production and distribution for crop production.

Seed certification

The Association of Official Seed Certifying Agencies (AOSCA) is the main organization for establishing standards for genetic purity, cultivar identity and standardization of seed certification in North America. This organization also includes some South American countries and New Zealand in addition to Canada and the USA.

The primary aim of seed certification is to check the crop from which the seed is produced and link this verification with agreed minimum standards of other important features of the seed lot. These include health, potential germination and mechanical purity. Thus the structure of a properly functioning seed certification scheme is complex. This is necessary in order to ensure that all the facets of seed quality are accounted for.

The generations of seed in a certification scheme are:

- *Pre-basic seed* – seed material at any generation between the parental material and basic seed.
- *Basic seed* – seed which has been produced by, or under the responsibility of the breeder and is intended for the production of certified seed. It is called 'basic seed' because it is the basis for certified seed and its production is the last stage that the breeder would normally be expected to closely supervise.
- *Certified seed* – this is the first generation of multiplication of basic seed and is intended for the production of vegetables as distinct from a further seed generation. In some agricultural crops there may be more than one generation between basic and certified seed, in which case the number of generations of multiplication after basic seed is stated (e.g. first or second generation).

Generally, many of the vegetable crop species have higher multiplication rates than the agricultural crops such as cereals. Vegetable species with relatively low multiplication rates include members of *Leguminosae* such as peas and beans. It is usually the multiplication rate of a species which determines whether, or not, further generations can be produced beyond the first multiplication from basic seed.

The success of a seed certification scheme depends on the collection and compilation of evidence from several aspects of seed production and quality control. Furthermore, it also relies on demands for certified seed from vegetable growers. Thus a successful certification scheme depends on the ability of a country to produce seed of high quality in sufficient quantities to meet market requirements. It is the culmination of successful seed programmes which over several or even many years have established the required infrastructure of a seed industry. If seed certification schemes in a country aim higher than the industry's current realistic capability, then not only will they fail but their shortcomings or failure will be detrimental to any future development of seed legislation or seed control in that country.

Standard seed is a category that contains seed which is declared by the supplier to be true to cultivar and purity, but is outside the controls of a certification scheme.

The Quality Declared Seed System

The Quality Declared Seed (QDS) system was introduced by FAO in 1993 and later revised and updated (FAO, 2006). The scheme has been designed for 'True Seed' and is of particular value and application in countries where there are insufficient resources, or lack of infrastructure for the establishment of highly developed seed monitoring systems, such as seed certification. The QDS system is of importance and value for:

- Seed purchased for emergency relief supplies.
- A reference scheme for these purposes, especially in international seed movement.
- The scheme can also be applied to assist other potential seed suppliers, for example farmers' groups and cooperatives, private farms and NGOs who enter into seed-supply activities for seed quality assurance.

The system has been designed to ensure that a country's existing seed quality control resources are used to maximum advantage. The concept of the QDS system is to enable farmers and growers to have access to seed material of a satisfactory standard. It also recognizes the role of extension services in the demonstration of improved seed to the farming community. QDS is not intended to be an alternative to, or in competition with a more developed seed quality control system, or duplicate the work of specialist organizations. The documentation of the system refers to *varieties* rather than *cultivars*. The scheme recognizes three types of varieties:

1. Varieties developed through conventional breeding technologies.
2. Local varieties that have evolved over a period of time under particular agroecological conditions and are well adapted to the conditions of that area. A local variety is sometimes called a 'land race' or an 'ecotype'.
3. Varieties developed through alternative plant breeding approaches, such as participatory plant breeding.

The legal framework of the system requires that a participating country establishes a list of cultivars eligible for inclusion and that participating seed producers are registered with the relevant national seed authority. The national authority is responsible for checking at least 10% of seed crops entered in the scheme and at least 10% of 'QDS' offered for sale in the country. The *Quality Declared Seed System* (FAO, 2006) provides crop-specific sections for some 82 crops, including eight food legumes and 35 vegetables; there is provision for adding further crop species. The crop-specific sections outline the requirements and obligations for each seed crop including: (i) facilities and equipment; (ii) land requirements; (iii) field standards; (iv) field inspections; and (v) seed quality standards.

Truth in labelling

This type of seed quality control does not have minimum standards but relies on the vendor making a statement as to the quality of seed offered for sale. The seed seller is obliged by law to state certain facts regarding quality. The statements made by the seller for individual seed lots are subject to random checking by the government seed control agency; most truth-in-labelling schemes are based on a 10% sampling. The required standards can vary from country to country, and indeed from crop to crop. The truth-in-labelling concept is a useful way of commencing seed quality control in the early stages of a developing seed industry.

Planting material

The majority of schemes developed for the control of seed quality do not apply to planting material for the vegetatively propagated vegetables. A lot of developments have taken place with some of the tree fruits, such as citrus and apple, with special attention to authenticity and uniformity of rootstocks, also disease indexing of scion and budding material, including cultivar authenticity. However, there has not been the same level of attention given to crops such as the tropical tubers and other vegetatively propagated vegetable species, especially crops of importance to subsistence farmers. FAO held an Expert Consultation in Lima, Peru in 1997 in conjunction with the International Potato Center (CIP). The object was to formulate protocols for reduction of viruses, other pathogens and pests transmitted by plant material. The Consultation also aimed to provide minimum standards for vegetatively propagated planting materials which could be implemented by plant propagators or extension workers at community level. The objective was to raise the overall quality of 'seed' (i.e. planting material) available to smallholders. The outcome was the publication of protocols for some 12 tropical crops including tuber and root vegetables, potato (*S. tuberosum*), banana and plantain (FAO, 2010).

These protocols and standards are designed to be complementary to the formal seed quality control systems but be appropriate for, and achievable by, its users; it takes national and local conditions into account. The scheme focuses on healthy planting material production at community and extension levels.

The principles of vegetative propagation embrace the following:

- Multiplication of plant material by non-sexual (vegetative) methods which produce cloned material – the cloned material, or propagules, from one selected plant normally have the same genetic makeup and would generally be expected to be uniform and normally have the same morphological characters as the original selected mother plant.
- Within the limits of the technology used, pathogens (including yield-reducing viruses) and pests can be controlled and 'clean' plants produced, although the level of control may be limited depending on the efficiency of the specific techniques and tests made.
- A higher level of multiplication and accuracy can be achieved in laboratories specializing in micropropagation incorporating specific pest and pathogen

indexing, including specific viruses. For example, in the case of potato (*S. tuberosum*), one explant produced by micropropagation can be subcultured and a fivefold increase can be achieved by each subculture. With rigorous and efficient screening of the plant material at the start of the programme and careful monitoring of the subsequent material produced, individual pests and pathogens can be virtually eliminated from the plant stock. The micropropagation and multiplication to produce potato planting material has been described by George (1986). Similar techniques and systems have been developed for other vegetable crop species, especially the tropical tubers.

It is to be hoped that the very 'clean' and 'true-to-type' plantlets from authentic micropropagation laboratories would be supplied to subsistence farmers at village level, for example via government, extension or provincial-led schemes and services.

In some countries, or regions within countries, local tissue culture laboratories are run by groups of farmers. It is especially important that these locally operated laboratories are able to maintain or increase the standards of trueness to type and freedom from viruses and other plantborne pathogens by having access to improved pathogen-indexed propagules derived from materials produced, tested and distributed from regional pathology and biotechnology laboratories. In this way there is a better opportunity to improve the overall qualities of vegetatively propagated plant material available to subsistence farmers.

Suitable macrovegetative propagation methods for some individual root and tuber vegetables are dealt with in Part 2 of this volume.

Types of Cultivars Widely Used in Vegetable Crop Production

Open-pollinated cultivars

An open-pollinated cultivar (sometimes referred to as an OP) consists of a population which is at genetic equilibrium based on a stable inter-pollinating population. This type of cultivar is normally both heterogeneous and heterozygous. An open-cultivated cultivar is normally maintained by mass selection of mother plants according to visual assessment (usually referred to as 'roguing'). In practice, in a field or plot for seed production, those plants which differ significantly from the 'cultivar description' are removed from the crop before producing pollen. The remainder of the plants are allowed to cross-pollinate. Approximately 20% of the observed off-types (rogues) are removed but this figure will depend on the observed degree of variation in the crop. A cross-pollinated crop, such as a cucurbit, is likely to show a higher degree of variation than a self-pollinated crop such as tomato, but the observed variation will also depend on the genetic quality of the stock seed used. For further details and information on seed production of a wide range of vegetables see George (2009) and for seed production of agricultural crops see Kelly (1988).

F₁ hybrids

This type of hybrid is now available in a wide range of vegetable species, and production of F_1 hybrids is by far the most widely used hybrid technique used by vegetable breeders. In practice each of the two parent lines are the result of inbreeding, and subsequent crossing of two inbred lines which have been maintained under the control or supervision of plant breeders and which are known to produce a desirable hybrid.

The advantages of F_1 hybrid cultivars include uniformity, increased vigour, earliness, higher yield and resistance to specific pests and pathogens, although these factors are not always all present in any one cultivar of a vegetable species. F_1 hybrids which are resistant to specific eelworm or pathogens can be of great value in crop rotations. Heat tolerance has been an important character in the development of F_1 Chinese cabbage for tropical areas.

Theoretically all the plants in an F_1 hybrid cultivar resemble each other exactly, but because some self-pollination of the female parent used in the cross may take place, some plants which are not F_1 hybrids may occur and are usually morphologically different and are generally referred to as 'sibs'. The term *sib* is used to indicate that this non-F_1 hybrid material has resulted from 'sister' or 'brother' plants cross- or self-pollinating with the female line.

Policies relating to production and use of F₁ hybrid cultivars

Hybrid seed is usually several times more expensive than open-pollinated seed for any given crop. This is because the hybrid seed producer has to recover the development investment and pass this cost on to the consumer; in addition there is an increased cost for the production of hybrid seed. In some crops the potential F_1 seed yield is lower per unit area than for open-pollinated crops of the same species. One main advantage of an F_1 hybrid to the organization developing and marketing it is the relative difficulty with which competitors can reproduce the cultivar; further protection is afforded with the introduction of Plant Breeders' Rights. Regardless of any legal constraints, a grower cannot successfully use seed saved from F_1 hybrids due to segregation in the following generation and the progeny will not have the same characters as the parental material. Growers should always be made aware of this fact.

It is sometimes suggested that F_1 hybrids provide commercial seed companies with a hold on the market because once growers have found an F_1 cultivar to be acceptable they will be obliged to continue using it in subsequent years. However, it is the author's observation that vegetable producers worldwide are very keen to adopt F_1 hybrids which have been demonstrated to be beneficial, provided that they can afford the extra seed cost. This assumes that there are no other constraints to maximize the cultivar's potential. None the less, it must be emphasized to subsistence farmers that hybrid cultivars' uniformity of maturity in crops from which there is only one harvest per plant, (such as cabbage), may not provide the continuity of supply which they expect from some open-pollinated cultivars. It should be stressed that individual hybrid cultivars should be judged on their agronomic and economic advantages to the individual grower.

Genetically modified organisms

The development of biotechnology related to crops and the introduction of its products via the seed industry have resulted in a new tool for the production of vegetable and staple food cultivars in addition to the more traditional methods of plant breeding outlined above.

As a result of biotechnologists developing procedures which involve the transfer of genetic material between otherwise incompatible species or genera, the transfer of desirable traits between unrelated species has become possible. The resulting materials from this application of the technology are referred to as genetically modified organisms frequently abbreviated to 'GMOs'; cultivars of this origin may also be referred as 'transgenics'. There are several different techniques which can be used by biotechnologists to produce GMOs, the techniques have been described by Cooper and MacLeod (1998).

Possible applications of genetic engineering include: (i) quality improvement of pest and pathogen resistance; (ii) quality of harvested crop; (iii) resistance to herbicides; (iv) production of male sterility (useful for hybrid production); (v) extension of postharvest storage potential (including shelf life); and (vi) nutritional value of a cultivar to the advantage of the consumer. Much of the earlier work relating to crops was led by international companies, but some countries have gone as far as releasing GMO material for further multiplication by farmers.

The development and potential of GMOs has become a very contentious issue. The reader is referred to the relevant publications in the 'Further Reading' section at the end of this chapter, including the *Cartagena Protocol on Biosafety to the Convention on Biological Diversity* (Anonymous, 2000).

However, although many organizations and individuals in the developed countries, such as in Europe appear to have a strong anti-GMO policy they do not always engage in the point of view of farmers and their dependents in the developing countries. This volume is not the appropriate forum to join in this ongoing GMO debate, but it is worth noting that the potential advantages of transgenics for the poorer farmers could include the following list of genetically modified (GM) crops that Africans would like to grow, as highlighted by Thomson (2007):

- insect-resistant African maize cultivars;
- crops resistant to African viruses;
- biofortified African crops;
- drought-tolerant crops; and
- maize resistant to the parasitic weed *Striga*.

When one looks at evidence from parallel examples with farming authorities and small farmers in developing countries in other regions of the world a similar picture is painted. One of the ironies of the debate, for or against GMOs, is that many of the potential advantages of appropriate GMOs would match the theoretical requirements of those who promote organic vegetable crop production (e.g. pest and/or pathogen resistance in cultivars to eliminate the use of chemical pesticides!).

The role of the International Seed Testing Association (ISTA)
in monitoring GMOs

The International Seed Testing Association (ISTA) has the important and vital role of the promotion of uniformity in seed testing and has appropriate committees for the different aspects of its activities. In evaluating seed lots for GM material the association's Task Force has the following roles:

- To produce additional seed testing rules for the detection, identification and quantification of GMO material in conventional seed lots.
- To organize proficiency testing for the detection of GMO material in conventional seed lots.
- To set up a platform for the exchange of information between seed testing laboratories.
- To identify stacked genes, publish performance test results and make documentation relating to testing for specific traits available.

The work and future programmes of the Task Force has been described by Haldemann (2008).

The potential sources of contamination of orthodox (non-GMO) crops from GMO crops which are flowering (e g. during seed production) could include:

- admixture with GMO material during harvesting, seed processing or packaging; usually referred to as *adventitious* material;
- genetically engineered male lines designed for hybrid production;
- glyphosate-tolerant cultivars or lines;
- any cultivar developed for increased shelf life, for example tomato cultivars (including those crops for marketing fresh, processing or storage);
- cultivars engineered for changes in starch composition, for example maize for agricultural outputs;
- cross-compatible material within the family *Cruciferae* which has been designed for agricultural or biofuel outlets; and
- undesirable or deleterious characters which have entered or contaminated other seed crops, weeds or garden plants which have a common period of anthesis.

Modification of Seed Shape, Size, Seed Pellets and Protective Treatments

Pellets

The use of pelleted seed facilitates the manual and mechanical handling of seeds which are either small or a difficult shape to sow. Precision drilling direct in the field or sowing in modules using pelleted seed can ensure a high degree of accuracy regarding seed placement. Individual seeds are encased in an inert material such as montmorillonite clay. The implications of seed pelleting with special reference to seed quality was reviewed by Tonkin (1979). The pelleting of seed is usually done by specialist companies using proprietary processes at the request of individual seed companies.

Coating

Seed coating, or film coating, is a technique by which additives such as pesticides, nutrients or nitrifying bacteria can be applied to the external surface (i.e. the testa) of the seed. But in contrast to pelleting the coating conforms to the seed's shape and does not normally significantly modify the seed's size. An air-suspension technique originally used in the pharmaceutical industry is used (Wurster, 1959). A film-forming polymer containing the required active ingredient is sprayed on to the seeds while they are suspended in a column of air which is either heated or unheated. A colouring agent is usually incorporated at the same time and the coating materials dry very quickly, resulting in free-flowing coated seeds. The coating of horticultural seed with films has been described by Robani (1994); Scott (1989) has discussed the effects of seed coatings and treatments on plant establishment.

Further Reading

Seed and vegetative propagation

Food and Agriculture Organization (FAO) (1998) *Restoring Farmers' Seed Systems in Disaster Situations.* Plant Production and Protection Paper 150. FAO, Rome, Italy.

Food and Agriculture Organization (FAO) (2006) *Quality Declared Seed System.* Plant Production and Protection Paper 185. FAO, Rome, Italy.

Food and Agriculture Organization (FAO) (2010) *Quality Declared Seed of Vegetatively Propagated Crops.* FAO Plant Production and Protection Paper. FAO, Rome, Italy.

George, R.A.T. (2009) *Vegetable Seed Production*, 3rd edn. CAB International, Wallingford, Oxon, UK.

Climate and soils

Jackson, I.J. (1977) *Climate, Water and Agriculture in the Tropics.* Longman, London.

Schjonning, P., Elmbolt, S. and Christensen, B.T. (2003) *Managing Soil Quality: Challenges in Modern Agriculture.* CAB International, Wallingford, Oxon, UK.

Weed control

DeVay, J.E., Stapleton, J.J. and Elmore, C.L. (eds) (1990) *Soil Solarization.* Food and Agriculture Organization (FAO) Plant Production and Protection Paper 109. FAO, Rome, Italy.

Upadhyaya, M.K. and Blackshaw, R.E. (2007) *Non-chemical Weed Management: Principles, Concepts and Technology.* CAB International, Wallingford, Oxon, UK.

GMOs

Anonymous (2000) *Cartagena Protocol on Biosafety to the Convention on Biological Diversity.* The Secretariat of The Convention on Biodiversity, World Trade Centre, Montreal, Canada.

Cooper, W. and MacLeod, J. (1998) An introduction to genetically modified crops. *Plant Varieties and Seeds* 11, 131–141.

Ferry, N. and Gatehouse, A. (eds) (2008) *Environmental Impact of Genetically Modified Crops.* CAB International, Wallingford, Oxon, UK.

Thomson, J. (2007) Genetically modified crops – good or bad for Africa? *Biologist* 54(3), 129–133.

Tzotzos, G.T. (1995) *Genetically Modified Organisms: a Guide to Biosafety.* CAB International, Wallingford, Oxon, UK.

4   Support for Farmers

It is essential that to alleviate poverty, poor health, malnutrition and significantly improve the lives of subsistence farmers and their families, they receive well-explained details of applicable research findings and technical advances as soon as the information becomes available. To achieve this, schemes need to be developed to support the farmers and communicate and disseminate the relevant information from research findings quickly and efficiently using extension programmes.

Development and Establishment of New Schemes to Support Subsistence Farmers

Any new scheme which is proposed for the direct benefit of subsistence farmers should be undertaken with their consultation and cooperation. It is essential for the eventual success of the enterprise, whether it be funded by government or an NGO, that farmers should take an active part in the planning and implementation stages. The plan of development should recognize that there should not be ongoing and continuous support for an undefined period when the development is up and running, but that the target farmers who are expected to benefit should gradually take charge and be responsible for the scheme's operation. In this way it should become self-sustaining. However, responsibilities for manning and operating seed regulations, research and extension services should remain with the governments.

Lines of communication for dissemination of information

Smallholder farmers are often at the end of a long chain of research procedures, their findings and ultimate application. It is essential that they receive well-explained details of applicable research findings and technical advances as soon as the information becomes available. All farmers are sceptical of recommendations based on findings not achieved under their own conditions and constraints. It must therefore be appreciated and understood that all new cultivars, developments and techniques should be demonstrated at village level.

The smallholder farmer should be provided with adaptive research covering directly relevant information and demonstrations where applicable. If a local or district station can be established it should initially concentrate on the following aspects of vegetable production:

- improved seed and planting materials;
- vegetable production techniques, including crop planning and continuity of supply;
- up-to-date crop protection techniques and methods which are potentially available to farmers, including IPM;
- on-farm crop storage; and
- marketing – this may be considered in the near future for many smallholder farmers' capabilities, but if retail and wholesale marketing structures are to be developed, or prevailing ones reinforced, they should eventually be self-financing. When individual farmers or groups of small farmers increase their self-sufficiency they will be in a better position to market produce and develop an income for themselves.

The urgent requirement for increased use of inputs in agriculture and crop production is an important policy issue because most of the farming and crop-production communities live in rural areas and are dependent on agriculture and crop production for at least part of their income. Historically, increases in productivity were achieved through the expansion of cultivated or planted area, but as population pressure increases there is less scope to do this; few developing countries have been able to keep pace with the food needs of growing populations and food imports are rising steeply. Gordon (2000) has produced a very succinct report on the need for improving smallholder access to purchased inputs in sub-Saharan Africa; however, the points made could also be largely applied to other rural and farming areas in the developing countries in Asia and South America. Gordon points out that much of Africa's agricultural production is located in vulnerable, low-potential areas, and even higher-potential lands are now showing signs of environmental degradation. Gordon's last finding that the reform of agricultural markets has left many farmers with poorer access to purchased inputs is more specific to sub-Saharan Africa.

Research

Research which can benefit smallholder farmers is conducted at three levels: (i) international; (ii) national; and (iii) local.

International research

This is done at the main international research stations which are generally subject- or crop-group orientated. The majority are funded and organized within the framework of the Consultative Group on International Agriculture Research (CGIAR). The names and locations of CGIAR institutes are given in Appendix 2. These institutes conduct fundamental and applied research; most have field stations in other countries (regional stations) where their findings can be further applied and passed on to farmers via regional, national and local networks. The current policy of CGIAR is for the individual institutes to formulate collective action to alleviate poverty while also focusing on farmers' current and predicted problems resulting from climate change.

National research

Most, if not all, developing countries have at least one national research station devoted to agriculture, including vegetable crop production. In some countries they are associated with a university campus or are a department or branch of the ministry of agriculture or its national equivalent.

Local research

Where a local or district vegetable station can be set up it should concentrate on improving the productivity of subsistence farmers. The experiments and trials should extrapolate the results or recommendations of national and international research institutes and organizations. These should include priority topics such as:

- trials and demonstrations of improved cultivars from seed or planting material;
- identification and control of pests and pathogens, and promotion of IPM;
- production and distribution of improved quality planting materials, for example virus-free and authenticated tropical tuber crops; and
- crop nutrition and irrigation in relation to local growing requirements.

Extension

The concept of *extension* is that farmers and growers can receive advice and information originally derived from research but that is of practical use and relevant to their available inputs, location and level of development. Extension programmes should include the results of adaptive research (i.e. research findings that can be adapted to local conditions) and include women's agricultural activities as this would create an improvement in household food security and nutrition of the family unit. FAO (1998) estimated that women only receive approximately 5% of the extension services received by men. It is very important that there be a realistic proportion of woman research staff and extension officers in the appropriate research and extension organizations.

Innovators and laggards

In the context of agricultural extension van den Ban and Hawkins (1988) stated that: 'An innovation is an idea, method, or object which is regarded as new by an individual, but which is not always the result of recent research.'

When evaluating the 'take-up rate', usually referred to as 'adoption', of practical and theoretical advice given to farmers, it should be appreciated that there are several levels of adoption among farmers. Many workers who have studied the problems and advice of horticultural extension and its adoption have tended to categorize the recipient farmers according to their rate of adoption. Diederen *et al.* (2003) suggested that farmers could be classified in respect of their ability or attitude to change or progress as:

- innovators;
- early adoptors; or
- laggards (adopters of mature technologies or non-adopters).

They found that structural characteristics (farm size, market position, solvency, age of farmer) explain the difference in adoption behaviour between innovators and early adopters on the one hand and laggards on the other. They also found that early adopters and innovators do not differ from each other regarding these structural characteristics. However, they appear to differ in behavioural characteristics: innovators make more use of external sources of information and they are more involved in the actual development of innovations.

Diffusion of information

A definition of diffusion in this context has been made by Rogers (2003) who defined it as 'The process by which an innovation is communicated through certain channels over time among the members of a social system.' There are several different systems of technology and information transfer such as the training and visit system, Agricultural Service Centres, farmer field schools, mass media, formation of groups and training and advice for young farmers.

The training and visit system

This system is especially effective for communities which are composed largely of smallholder vegetable producers. The concept is that initially a small number of village farmers are each asked to act as *contact farmer* in their community, possibly only one per community.

The district or regional research organization holds regular 'hands-on' training sessions for appointed, trained *village extension workers*. During the course of these regular meetings and demonstrations the village extension workers are briefed on topical problems and current work that should be requiring farmers' attention.

Each contact farmer is visited on a regular and scheduled timetable. During the visit the contact farmer is given the topical advice by one of the trained village extension workers who has attended the fortnightly training session. The regular meeting between the contact farmer and the village extension worker can deal with current practical work and also future crop planning. Typical topics may include cultivar choice, updating agronomic practices, crop nutrition and topical pest and pathogen control.

This system can be developed as a two-way process for identification of current agronomic problems at village level and for them to be reported back to the parent research organization.

Agricultural service centres

The official title for these centres may vary from one country to another but the broad objective of their purpose is to remove constraints on farmers' progress and required inputs. The service centre may be responsible for sale and distribution of improved seed of appropriate cultivars. The distribution of fertilizers can be done from the centres along with other material requirements. The policy of pricing materials sold to farmers can be a delicate situation if it is likely to deter the development of private-sector sales. Conversely, some commercial seed organizations may not be reaching farmers in areas which are difficult to access, either because of terrain or distance. Some centres may have sufficient staff and land for demonstration vegetable plots and animal husbandry. Transplants derived from virus-free stock such as the tropical tubers may be distributed from the centres which can work in conjunction with specialist propagation units producing pest- and pathogen-indexed planting materials.

In smaller communities representatives from the centre can attend a weekly growers' market and have seed and transplants of recommended cultivars available for sale.

Farmer field schools

Farmer field schools are sometimes referred to as 'participatory technology development'. The objective of a farmer field school is to encourage groups of farmers from the same farming area to make decisions as a result of participating in group discussions. The background or basis of the group discussions is to encourage participants to consider their decision making and their individual approach to crop production and/or animal husbandry in the background of their individual farms. This is often referred to as the *agroecology* of their farms. During the group meetings, which are very 'hands on', farmers are encouraged and stimulated to organize their own farms and also contribute to the communal affairs of their home locations. The long-term results would be expected to include:

- adoption of improved seed and planting materials of suitable cultivars;
- improved capability to observe and make diagnoses of problems and disorders as they arise;

- farmers become better informed and observant smallholders with a fundamental understanding and comprehension of horticultural science as it relates to their own activities; and
- improved interchange of information and topical items at village level.

The farmer field school concept has been applied in several countries, including Kenya and Sri Lanka. A useful evaluation of farmer field schools has been made by Tripp *et al.* (2009).

Mass media

There are several media systems which are used to provide farmers and growers with research results and regular updates on topics related to tropical vegetable production and its associated activities. Quite often there is a mixture, or blend, of agronomic, nutritional and health advice. The advantages of this are that topics appeal to a wider audience or readership, although the agronomic 'take-home' message should remain factual and 'undiluted'. It is important that the disseminators of the information use the most appropriate approach to reach the target 'audience'. The potential methods include: (i) television; (ii) radio; (iii) newspapers; (iv) newsletters; and (v) growers' notes (either provided for each farmer, or passed to the media for publication).

Formation of groups

Informal groups of smallholder or subsistence farmers may form so that the farmers can share information and skills. The groups can be formed on farmers' own initiative or as a result of external encouragement.

Training and advice for young farmers

This can be a useful strategy along with the use of *school gardens*. When one looks through the literature there seems to be conflicting ideas as to the importance of younger persons being innovators. However, there seems to be a general consensus of opinion that early lessons in crop production for the younger generation can help develop the next generation of crop producers. This appears to influence the young and 'up-and-coming' growers and can also indirectly influence their elders. Figure 4.1 illustrates schoolboys receiving practical instruction in seed sowing and the importance of improved quality seed.

Influence of Farmers on Development

There is an increasing recognition and need for the encouragement and establishment of direct links between research workers and farmers in developing countries (Eponou, 1996). The improved liaison between the end user and the researcher (including plant breeders) offers an excellent opportunity for farmers and their dependents to express their views and influence the outcome of related research, and also plant breeding programmes and their resulting cultivars. The findings and results should always be demonstrated to farmers at village level under farmers' own conditions.

Fig. 4.1.   Hands-on instruction of seed sowing and use of quality seed.

Participatory research and development

The concept of participatory research and development is that farmers are able to give a valid and pertinent opinion on developments that will affect them directly. Thus the participation of farmers down to village level has the possibilities of embracing any, or all, of the following: (i) growers; (ii) consumers; (iii) marketers; (iv) processors; (v) policy makers; and (vi) nutritional and health experts. Where required or appropriate, specialists in food security or gender issues should also be involved.

Participatory plant breeding

Participatory Plant Breeding (PPB) is the term used for plant breeding programmes in which the developing lines or plant materials, while still under the jurisdiction of the breeder, are evaluated on a site, or sites, on which the breeding lines are cultivated under local growing conditions. It has the objective of encouraging local growers and communities to participate in the selection process and for them to have an influence on the development of cultivars for their specific agronomic conditions and product requirements. For example plant breeders and farmers have jointly contributed to the evaluation of bean cultivars in Rwanda (Anonymous, 1995). It is also important for plant breeders to appreciate that farmers often have a secondary use for a plant species and its products in addition to its food value.

The essential advantages of PPB are that developing breeding lines are evaluated under the future user's conditions and that all requirements are taken into account during the selection and cultivar development process.

This system of evaluation and cultivar development is used, where appropriate, by many plant breeders in both public and private organizations. For example the International Center for Agricultural Research in the Dry Areas (ICARDA) ensures that the local growers' cultural conditions and requirements are taken into consideration. PPB may also be applied by government breeders and commercial seed companies in breeding programmes designed to develop cultivars for organic vegetable production or any other specified agronomic requirement.

Further Reading

Diederen, P., Meijl, H.V., Wolters, A. and Bijak, K. (2003) Innovation adoption in agriculture: innovators, early adopters and laggards. *Cahiers d'Economie et Sociologie Rurals* No. 67, 30–50.

Food and Agriculture Organization (FAO) (1998) *Rural Women and Food Security: Current Situation and Perspectives.* FAO, Rome, Italy.

Food and Agriculture Organization (FAO) (2001) *Incorporating Nutrition Considerations into Agricultural Research Plans and Programmes.* FAO, Rome, Italy.

Gordon, A. (2000) *Improving Smallholder Access to Purchased Inputs in Sub-Saharan Africa.* Natural Resources Institute Policy Series 7. Natural Resources Institute, Chatham Maritime, UK.

Rogers, E.M. (2003) *Diffusion of Innovations,* Free Press, New York.

Tripp, R., Wijeratne, M. and Piyadasa, V.H. (2009) What should we expect from Farm Field School? A Sri Lanka case study. *World Development* 33(10), 1705–1720.

van den Ban, A.W. and Hawkins, H.S. (1988) *Agricultural Extension.* Longman Scientific and Technical, Harlow, UK.

The CABI Bioscience Centres produce a range of *CABI Discovery Learning Manuals* on specific topics for use by organizers and trainers who are planning or implementing farmer training courses. The materials cover a wide range of farmer-related topics for a range of geographical locations. There are centres in Kenya, the Caribbean and Pakistan, in addition to the UK and Switzerland. For the available titles contact: www.cabi-bioscience.org

5 Crop Preparation and Management

Equipment and Tools

The tools for small farming operations such as land preparation, sowing, planting out, weed control and harvesting can be divided into *mechanized* and *hand tools*.

Mechanized tools

The subsistence farmer does not normally have the financial capability or backing to invest in mechanical aids, certainly not at the outset. Sufficient income should be assured to maintain and replace machinery that has been donated through projects or development programmes (this also applies to mechanized tools given to groups of farmers) otherwise the mechanical tools will fall into disrepair resulting from lack of spares or replacement parts.

Where the situation allows, the most useful and versatile piece of equipment is the smaller rotary cultivator although it is not normally recommended for stony or rock-strewn soils. These cultivators can have a wide range of attachments and, depending on cost and the manufacturer's model attachments can provide: (i) cultivators (rotary cultivating); (ii) ridgers; (iii) hoes; (iv) a cutter bar (e.g. for cutting crop remains or fodder crops); and (v) when the machine is stationary the engine can be used for 'power take off' for spraying and pumping accessories. There are some very versatile models on the market.

Hand tools

These may be classified according to the type of work that can be achieved.

Land clearance

The hand tools most frequently used for this purpose are the mattock and the pickaxe. The *mattock* has a head with two arms, one like an *adze*, at right angles to the handle or shaft, the other like a small axe, in line with the shaft; it is a very useful tool for clearing rough ground, especially scrub and for removing roots. In some circumstances it is used for rough digging of heavy soils. The mattock may also be used for opening up shallow trenches or furrows. The *pickaxe* (also called a *pick*) has an anchor-like curved head, one end is pointed and the other is chisel like. This tool is used for clearing rocky sites, making trenches and furrows in hard or rocky ground. The *machete, panga* or *cutlass* is a large knife with a very sharp metal blade approximately 30 cm long with a wooden handle to balance it. This implement has a range of names and design of blade shape in different areas of the world. It is invaluable for clearing overgrown plots or virgin bush, in these cases it is used in a 'slashing' motion, keeping the other hand well out of the way (the uninitiated are well advised to keep their other hand behind their back!).

Digging

There are many types of spade used in the world. The main purpose is for turning over the soil in parts of the plot which are vacant, or for digging in plant debris or other organic materials. It is important that *spades* or *forks* are of suitable weight and size to cope with soils, which are either heavy or light. If the soil surface is relatively free of debris a spade is very suitable. However, if the amount of material to be incorporated into the soil impedes the digging rate and efficiency a fork is likely to make the task easier and more efficient. There are several local designs of local digging implements; Fig. 5.1 illustrates two such tools used by smallholder women farmers in Kenya.

Transplanting

Transplanting is the planting out from seedbeds or containers. Generally, the tool used will depend on whether the transplant has been raised in a seedbed or grown on in a container. The *dibber* is the tool usually used for transplanting plants with loose roots (as lifted from a seedbed). A hole is made with the dibber; the depth of the hole is in accord with the length of root and the optimum point where the lower stem is to be at finished soil level. The plant is placed in the hole and then it can either be firmed in by hand or by inserting the dibber close to the original planting hole; with practice the latter is more efficient. A *trowel* is usually the best choice of tool for hand planting material from containers such as plugs, pots or boxes. A hole is made at the planting position and the soil pulled forward towards the worker, holding it back with the trowel while at the same time inserting the plant into the hole, the trowel is taken out of the soil and the soil pushed back and firmed in. As with developing any skill, with practice the task can be completed in minimal time.

The firming in of the transplant is equally important whichever method is used. A quick check, or *post-planting test*, is to hold the tip of a leaf between index finger and thumb immediately after planting and gently pull, if the plant

Fig. 5.1. Two digging implements used in Kenya, single adze-type blade known as a 'jembe' and a three-tined implement known as a 'forked jembe'. Photograph courtesy of The Real IPM Company (K) Ltd (www.realipm.com).

has been sufficiently firmed in the soil the leaf tears which clearly demonstrates that the firming has been sufficient.

Other preharvesting operations

Shovels are similar to spades but usually have a broader blade with upturned sides. They are unsuitable for digging but ideal for mixing composts, moving, loading or unloading finer-graded materials such as sand, wood ash and fine particle composts.

Rakes may have metal or wooden tines. *Metal tined rakes* are usually used for the preparation of seedbeds or final planting quarters for transplants. Raking in the final stages of site preparation is usually the best method to obtain a fine tilth for seed sowing. *Wooden tined rakes* are generally wider than the metal-tined type and are used for raking off loose plant debris at the end of a crop, or for gathering up mown fodder such as hay.

There are essentially of three types of hoe. The *Dutch hoe*, with its long handle with the blade on the end cuts weeds off just below soil level. The *draw hoe* is a very versatile tool, useful for weeding, earthing up crops growing in ridges or taking out a seed drill against a taught line at soil level, while drawing the hoe backwards. The *onion hoe* is a short-handled 'half-moon' hoe used for the first stage of onion harvesting, it is used to cut the maturing bulb's roots below the bulb prior to further field drying of the loosened bulbs.

There are many types of *sprayers* on the market, mainly varying in their capacity and throughput. The most easily handled and versatile is the *knapsack sprayer*.

Harvesting

There is a wide range of morphological types of vegetables at their harvest time. Many of the crops such as pot herbs, tomatoes and legumes are picked by hand. In the case of root and tuber crops produced in ridges, the tubers or roots are lifted with a digging fork or a locally improvised implement usually referred to as a '*stick*' or '*lifting stick*'. This implement, fashioned from natural wood, is approximately 40 cm long and is simply pushed into the soil at an angle of approximately 45° just below where the tubers are anticipated to be, then the top of the stick is levered down and this action lifts the tubers out of the soil. It can be imagined that this tool can be very effective in crops grown in ridges or raised beds. The digging fork is also a widely used tool for this purpose, but care must be taken to avoid mechanical damage to the harvestable crop as this would reduce its value and storage life. The machete, panga or cutlass, described above, is very useful for harvesting headed crops such as cabbage otherwise a sharp knife can be used.

Field and Crop Management

Crops may be grown one at a time in rotation (crop rotation) or two or more crops may be cultivated together (referred to as intercropping, companion planting or mixed planting).

Crop rotation

Vegetable crop production is relatively intense and continuous which is by necessity on small farms where all the available space is usually cultivated to maximize yields.

The objectives of crop rotation are to maintain and, as far as possible, to improve soil fertility. Fertility in this context includes soil nutrients and soil physical structure. This objective may be achieved by:

- including a legume crop in the rotation sequence;
- ploughing or digging crop residues into the soil;
- sowing a fodder crop the remains of which are incorporated in the soil;
- sowing a grazing crop which will be grazed *in situ* thus providing both a break from vegetables and also the advantage of additional organic material from the livestock and subsequent turning in of the crop remains; and
- growing a sequence of unrelated vegetable species using different soil horizons (i.e. crops with shallow or deep root systems).

Other advantages of growing crops in rotation are:

- to reduce infestation of soilborne pests and pathogens (e.g. nematodes, Verticillium and Fusarium wilts) by growing a sequence of unrelated crops with no common soilborne pests or pathogens. Other advantages can be obtained by following vegetables by so called 'trap crops' such as groundnut. A 'non-host' crop such as *Tagetes* spp. is reputed to be a non-host of *Meloidogyne* spp. and is used as a break between susceptible crop species; and
- for farm management, by organizing a satisfactory work schedule through all the consecutive growing seasons and making optimum use of available land, man power and other available facilities.

Intercropping, companion planting, mixed planting

These three terms embrace the traditional and flexible sequences of cultivating more than one crop on the same site at the same time but exclude the practice of producing different crop species in separate rows or blocks of pure stands on the same plot. Figure 5.2 illustrates the intercropping of cauliflower, sweetcorn and tomato in Indonesia.

For many growers the term *intercropping* infers growing a faster maturing crop between the rows of a slower developing crop, so that the faster maturing species

Fig. 5.2.   Intercropping sweetcorn with cauliflower and tomato.

is harvestable before the slower vegetable species 'fills the rows' (i.e. making the most economic use of the inter-row space in a short time without competing with the slower developing crop). However, the same term can also mean growing two or more crops in the same plot, but in a spatial arrangement or pattern which is fairly symmetrical; the plant arrangement can take into account the optimum spacing for each individual species in the arrangement. Examples of short-term crops include: okra, shorter cultivars of *Amaranthus* and hyacinth bean. Examples of longer term crops include: tomato, onion, peppers, cabbage and sorghum.

Another form of intercropping is the use of one plant as a support for another plant. For example in a developing stand of maize plants a seed of a climbing bean cultivar is sown alongside each maize plant and the maize plants provide support for the climbing beans. This technique requires some local experience and cultivar information to obtain the best dual crop sowing and harvest timings and yields, but is widely used by subsistence farmers.

The *multi-storey smallholder cropping system* for plantation crops is usually a three or four tier system. Awoke (1997) interviewed smallholder coffee growers in Ethiopia and found that there are modifications between the western, southern and eastern zones. However, the basic system described by Awoke is based on several crops growing on the same piece of land using successive canopy layers to provide shade for the crop below it. The top storey is composed of shade trees grown for timber production; the coffee forms the next layer down along with taller food crops such as maize, sorghum and legumes; while the ground level crops consist of vegetables, tropical root crops and spices. The crop mixtures as outlined above have continued to be widely adopted by subsistence farmers. Generally, intercropping in the tropics is more suited to crops grown in raised or sunken beds than ridges.

The potential advantages of intercropping are:

- Satisfactory competition with weeds.
- Making a more uniform and even labour input pattern over the cropping periods.
- An insurance of food supply – it has long been believed that intercropping provides an assurance over failure of a single crop, in the sense that if one crop fails as a result of pests, pathogens or adverse weather, the companion crop or crops will provide a food supply.
- The potential for a higher cash return per unit area – researchers in Africa during the 1950s and 1960s generally found that appropriate intercropping gave a higher cash return per unit area than growing the two crops in separate plots. Later work in Nigeria (IITA, 1974, 1975) and in India (ICRISAT, 1974) demonstrated that maize with soybeans and also mung bean, sweet potato, or groundnut with irrigation yielded a 30–50% increase compared with pure stands of each crop. It is important to bear in mind that the optimum population of each species is less than the population density would be in a single species stand.

Mono-cropping is usually taken to imply that the same vegetable crop is continually grown on the same site, season after season. It requires very efficient control of soilborne pests and pathogens, using appropriate soil sterilants and/or partial soil sterilization by heat. It is a system usually adopted for the production of high value crops where the crop's market value can justify the relatively high cost of soil or substrate treatment; mono-cropping is therefore unlikely to be used by subsistence farmers.

Weed control

A weed may be defined as a 'plant in the wrong place', but in the tropics it may also be an indigenous species, although in theory considered as weeds some indigenous species are of culinary importance and nutritional value. It is therefore important not to condone all 'weeds' in the tropics out of hand because in some situations, such as mixed crop stands, they may be considered as essential sources of nutrients; however if they are likely to reduce crop yield or be host plants of pests and pathogens then they should be regarded as weeds.

On sites where there is a known threat or risk of runoff there can be a case for allowing weeds to establish so that the soil cover is increased.

Generally weeds may be thought of as undesirable because they may:

- Compete with the vegetable crop for nutrients, water and photosynthetic light. The competition will start very early in the crop's life under tropical conditions.
- Be alternative hosts for crop plant pests and pathogens, including viruses.
- Impede crop harvest.
- Impede air circulation within the crop canopy and thus provide a higher humidity for the germination and growth of fungal and bacterial pathogens.
- Be important parasites on the crop plants. Economically important parasitic weeds include: *Orobanche aegyptiaca* Pers., *Orobanche ramose* L., *Orobanche cernua* Loefl., *Cuscuta campestris* Yuncker (see Fig. 5.3), *Striga asiatica* (L.) O. Kuntze syn. *S. lutea* Lour. and other *Striga* species.

Vegetable plants which have been transplanted start to form their own canopy before weeds have germinated. The significance of the effect of weeds may, to some extent, depend on the crop species, especially the weed species' time to germinate and emerge, also its relative spread over the soil and formation of a competing leaf canopy. For example, in onions and related species, germination is relatively slow and the *Allium* spp. do not form a significant crop canopy; furthermore an onion plant's leaf number is low and its spread is relatively sparse compared with its likely weed competitors which may be rosette forming. On the contrary, some of the legumes, such as *Phaseolus* spp. germinate relatively quickly and soon form their own canopy, thus suppressing weed development from an early stage of crop establishment.

Sources of weeds in cultivated sites are:

- Dormant seeds in the soil.
- Weed seeds and plant materials derived from previous plant debris in the soil. Those derived from a previous crop are often referred to as 'volunteers' or 'ground keepers'.
- Weed seeds and weed plant material introduced from mulching, compost and organic manures.

Fig. 5.3. *Cuscuta* sp. on a leafy vegetable.

- Weed seeds and weed plant material introduced via livestock, footwear, tools, vehicle wheels and cultivating machines.
- Weeds which have been allowed to produce seed before they are controlled.
- Impure crop seed lots; this is a particular dangerous source for noxious and parasitic weed seeds, such as *Cuscuta* and *Orobanche* spp. which are serious parasitic weeds; also *Striga* species ('witch weeds') which are semi-parasitic. Figure 5.3 illustrates the threadlike growth of dodder on a green-leafed crop. Farmers' own saved seed or seed stocks which have not been subject to satisfactory seed quality control may contain undesirable weed seed, including seeds of noxious weeds and great care should be taken when procuring seed lots from unreliable or untraceable origins.

Stale seedbed technique

This is a technique that allows weed seeds to germinate in a prepared seedbed before the crop seed is sown. As soon as the first flush of weeds show, the soil is surface tilled, for example by shallow hoeing. Other alternatives to hoeing include flaming or an application of the herbicide glyphosate. The stale seedbed technique may also be referred to as 'the flush system' or 'false seedbed' method of weed control. The vegetable crop is then sown (or planted) with the minimum disturbance to the seedbed. This is a long-standing traditional system having been used for many years before the advent of herbicides or organic growing.

Growth Media

The substrate used to produce transplants is referred to as either growth media or container media; these terms are used when the transplants are raised in any one of a range of containers such as boxes, plastic pots, peat or other suitable materials including a range of proprietary containers and multi-cell-systems. If seedlings for transplants are to be raised in beds, similar growth-media formulations should, as far as is possible, be used for the substrate.

Compost ingredients and preparation

The subsistence farmer is at a major disadvantage for the availability of suitable ingredients to prepare composts. The farmer is very unlikely to be able to purchase ready-mixed proprietary composts that are available in more developed horticultural areas and is also not very likely be able to access the most suitable basic materials. Many of the potting and container composts developed from about the mid-20th century contained a proportion of peat in their formulations. But pressure from conservationists and environmentalists to reduce the decline of peatland stimulated the use of alternative organic materials, especially recycled ones. There are now several approaches to use recycled and local materials in horticulture from composted organic materials and by-products from other areas of agriculture and forestry such as waste mast from vine seeds, shredded tree bark, rice husks, coir waste and coco peat. It has now become important that local research is accelerated to determine the best composts that can be made from these materials.

The main requirements of a satisfactory container media, or compost, are:

- Ability to retain water which is available to the plant; the soluble salt content should not inhibit water uptake.
- Sufficient porosity for free air movement allowing interchange of oxygen and carbon dioxide between the air and the roots.
- Presence of available nutrients required by the plant for its duration in the container. This is usually achieved by incorporating the appropriate organic and/or inorganic materials during compost preparation, also by subsequent top dressing, liquid feeding, foliar feeding or a combination of these methods. Attention must be given for the provision of the necessary macro- and micronutrients when relatively inert materials are used for the compost because many of these nutrients are essential for the plants in their early and following stages.

Production of container compost

The optimum mixture of container compost for seed sowing is equal parts of partially sterilized soil and sand. When prepared as potting compost the ratio is 2:1 soil to sand, although other mixtures based on vermiculite, perlite and peat are used.

The soil component

Ideally the soil should have been partially sterilized. There are propriety soil sterilizers on the market which are capable of partial sterilization of soils. The term *partial sterilization* is used because the objective is to control weed seeds and harmful pests, pathogens and bacteria but not significantly reduce the population of beneficial bacteria which are able to survive temperatures up to 82°C.

The most widely used method is the application of heat and there are several types of equipment designed for the operation including electric and steam soil sterilizers, however, the cost of investment and/or lack of appropriate energy sources usually make these unavailable to the subsistence farmer. Chemical methods have been used, especially fumigation by methyl bromide, but this chemical has now been generally withdrawn in many countries because of its high toxicity to human life and toxic soil residues which enter the food chain.

With regards to heat treatment the soil to be treated should be dry and have a fine tilth. The heat will take longer to penetrate clods and heat is wasted if the soil is too wet. Many farmers heat the untreated soil in a supported trough with the heat source underneath. The heat front moves up through the soil from the bottom to the top. Soil temperatures are measured with a soil thermometer; the heat source is stopped when the required temperature has reached the surface. Efficiency of the process is increased if the soil surface is covered with a tarpaulin throughout the operation. The treated soil is removed from the heating container when it has cooled down. It must be emphasized that the partially sterilized soil must be handled with clean tools and equipment and stored on a clean surface, where it will not become contaminated by passing footwear, livestock, barrow wheels or any other source of contamination.

The sand component

River sand is frequently used if it is obtainable. Ideally fine particle sands should be avoided as they tend, in time, to form a hard cap when mixed with soils with a high clay content. If sea sand has to be used, it must be very well washed to remove residual salts.

Production of Transplants

Seedlings for transplant production are either sown in prepared seedbeds or containers, such as seed trays ('flats') or in modules. Modular production reduces root disturbance at the planting out stage. Examples of modular transplant production are illustrated in Fig. 5.4.

Fig. 5.4. Examples of modular containers for the production of transplants.

The advantages of transplant production in containers or seedbeds are:

- Less seed is used compared with direct sowing in the field (up to five times more seed may be used to produce one plant when direct sowing in the final quarters). Smaller amounts of required seed can enable the farmer to purchase better quality seed and make a significant improvement to the final harvest.
- Advantage can be taken of pelleted or coated seeds for ease of handling and more precise placement in containers or a single module as an individual seed can be placed at each sowing station.
- Seed and seedlings can be better protected from adverse conditions, including sun, wind, heavy rain, control of pests and pathogens in the crop's early and formative stages.
- Transplants can be grown on in a nursery during the dry season ready for transferring to the field when the rains arrive. The same concept can be applied in areas which experience cold seasons; transplants are raised under protection and then hardened-off prior to the arrival of warmer conditions.
- Transplants can provide a quick turn round to replace expired field crops, thus making a significant contribution to land use and continuity of supply. A crop which may take up to 8 weeks to grow before its transplanting stage is reached will allow an extended cropping time for the crop which it is replacing in the field.
- Seedling disturbance during transport to final planting areas is minimized.
- There is less 'transplant shock' because of minimum root disturbance when planted out.
- There is improved security and protection from rodents, feral and domestic animals.

- In a crop which is intended for on-farm seed production, transplants provide an early opportunity for roguing out off-types or making selections based on young plant characters before or during planting out.

Disadvantages of transplant production in containers or seedbeds, with special reference to subsistence farmers, include:

- cost of containers;
- availability of suitable constituents for preparation of the compost; and
- suitable site and facilities for producing the seedlings.

Production of transplants in seedbeds

Seedbeds are best laid out so that workers can reach the centre of the bed from either side and there is sufficient path width for access with various implements such as barrows, sprayers and irrigation equipment. Figure 5.5 shows vegetable seedbeds in a vegetable plant nursery in Cape Verde.

Bed preparation
The beds are best prepared well in advance of sowing so that the stale seedbed technique can be applied ahead of sowing. Initially the bed is dug and well-composted organic materials are incorporated; fresh livestock or fresh poultry manure should be avoided as either may cause ammonia or nitrite toxicity to young seedlings. After firming down by treading (depending on the local soil conditions) any fertilizers required for early plant development, especially phosphorus,

Fig. 5.5. A nursery bed area for the production of transplants. Note the use of straw mulch on the seedbed in the foreground to keep the soil cooler during germination.

Fig. 5.6.   Shading of seedling beds using bamboo hoops and palm foliage.

are added and raked in. It is usual to apply the main nitrogenous and potassium sources in stages after planting out. If overhead shading or other environmental protection is planned then it should be set up at this stage, providing it will not impede the sowing process. Figure 5.6 illustrates the provision of shade to protect the vulnerable seedlings from the direct sun; as the young plants develop the palms over the hooped frames can be gradually thinned out so as to 'harden-off' the plants prior to transplanting in the open.

Seed sowing

Seed is sown across the bed in drills 15–30 cm apart, the depth of the drill should depend on the seed size. As a rule of thumb seeds are sown at a depth of approximately four times their diameter, it may be necessary to modify this depending on the depth of any mulching material which will be added after filling in the drills and lightly firming the bed after sowing.

Soil Solarization

'Soil solarization is a hydrothermal process that occurs in moist soil which is covered by plastic film and exposed to sunlight during the warm months' (Katan and De Vay, 1991). This quotation provides a clear and comprehensive definition of soil solarization. The technique is of particular use where there are suitable periods of solar radiation, it is adopted in areas where farmers either cannot afford or have facilities for other methods of fumigation and partial sterilization. With the increasing policy to use less pesticides and less fossil-derived fuel it is ideal. The practical application for the subsistence farmer is that the energy source comes from the sun and the only cost is the plastic film and labour. The operation is normally done during the summer or start of the hot dry season, depending on local climatic conditions. The technique may also be acceptable for organic vegetable production requirements. Soil solarization may be expected to control the soilborne wilt pathogens (e.g. *Verticillium* and *Fusarium* spp.), root rots, eelworm, weed propagules and dormant weed seed. An interesting study of the thermal death points of some weed species' seed has been made by Dahlquist *et al.* (2007).

The soil solarization method is carried out using the following sequence of steps:

1. The area, or bed, to be treated is initially tilled in order to break up clods and large pieces of plant debris which may remain from the previous crop.

2. The surface is then carefully raked to produce a smooth surface so that the soil has uniform contact with the plastic which will cover the surface. Any sharp stones should be removed from the surface at this stage.

3. Dry soils should be irrigated after preparation; this is because the heat transfer takes place as a result of conduction down the soil's profile which is more efficient when the soil is wet.

4. The surface area of soil to be treated is covered with a thin transparent polyethylene film (25 µm), usually at the start of the dry season. These thin films can usually survive for approximately 10 weeks before ultraviolet (UV) light degradation commences. It should be noted that the thinner the film the lower the cost per unit area of soil covered, although thinner films are at greater risk of damage.

5. The polythene sheet is laid over the bed in such a way that it remains taught and in close contact with the soil surface, meanwhile the free edges are dug into shallow furrows and firmed down. It may be necessary to place rounded stones or bags of soil on the top of the sheet to avoid billowing. Any adjustments which require walking on the plastic surface should be done in bare or stocking feet.

6. The plastic should remain in position for approximately 4–6 weeks; ideally the soil temperature beneath the plastic should be at least 85°C. The temperature can be confirmed with vertical placement of one or more soil thermometers.

Further Reading

Flegmann, A.W. and George, R.A.T. (1975) *Soils and Other Growth Media*. Macmillan, London.

Food and Agriculture Organization (FAO) (1991) *Major Weeds of the Near East*. FAO Plant Production and Protection Paper 104. FAO, Rome, Italy.

for weed seed thermal death. *Weed Science* 55, 619–625.

Food and Agriculture Organization (FAO) (1991) *Soil Solarization*. Plant Production and Protection Paper 109. FAO, Rome, Italy.

Katan, J. and De Vay, J.E. (1991) *Soil Solarization*. CRC Press, Florida, USA.

Soil solarization

Dahlquist, R.M., Prather, T.S. and Stapleton, J.J. (2007) Time and temperature requirements

6 Reducing Pre- and Postharvest Losses and Marketing Surpluses

Pre- and postharvest losses may result from pests, pathogens and physiological disorders. This chapter discusses ways in which such losses can be reduced using for example pesticides and Integrated Pest Management (IPM) and continues with an outline of the type of markets that develop in order to sell any surplus produce left after the subsistence farmer has met the food needs of his or her immediate dependents.

Pest and Pathogen Control

The spread of pests and pathogens

The pests and pathogens which are of concern to vegetable producers can be spread in several different ways, these include:

- Imported plant and seed materials (i.e. imported from other countries or regions).
- Locally produced plant and seed materials, including neighbours' own-produced (on-farm) plant materials. In the case of seed transmission the pest or pathogen may be on or in the seed, or as an admixture in the seed lot.
- Wind, soil, irrigation water and rain splash.
- Workers' footwear, clothing, knives, secateurs, hand tools, barrows and machinery.
- Animal coats and feet.
- Insects and weeds.

Definition of a pesticide

From the point of view of agricultural and horticultural crop producers a *pesticide* is a substance for the control of insects and other pests of plants (during their production and storage). It is necessary to clarify this statement by pointing out that the use of the word *pest* in this definition, in more recent and in current times includes:

- insects controlled by insecticides;
- aphids controlled by aphicides;
- mites (e.g. red spider mites) controlled by acaricides;
- nematodes (eelworms) controlled by nematicides;
- fungi controlled by fungicides;
- bacteria controlled by bactericides; and
- weeds controlled by herbicides (sometimes referred to as weedicides).

Growth regulators are regarded as 'pesticides' in the context of the above definition. Soilborne pests, pathogens, weed seeds and weed propagules *in situ*, are controlled by soil sterilants (some soil sterilants have a narrow spectrum while others have a wide spectrum of control).

Availability, suitability and approval of pesticides

There is a very wide range of proprietary pesticides available globally. It is usual practice in each country for individual materials to be approved as a specific control for any given individual or range of pest/s or pathogen/s. The approving authority in each country is normally a department within a ministry of agriculture, its equivalent or a neutral agency approved and appointed by the national or regional government with no direct links to pesticide manufacturers, their distributors or retailers. It is vital that the approval system has complete transparency.

Generally the manufacturer seeking approval for a specific product is required to submit results of field trials and laboratory tests; these would usually be expected to include aspects of toxicity for both plants and workers as well as effectiveness for the purpose for which the approval is being sought. Other evaluations would include the material's possible deleterious effect on: (i) public health; (ii) environmental issues; and (iii) water supply, water courses and groundwater. The overall evaluations would usually be made by a team comprising agronomists, toxicologists and occupational physicians. This team would, for example, reject and ban outright a submitted material which has been found to be a potential cause of carcinogens, mutagens, disruption of endocrines or reproductive disorders. Other less drastic dangers would also be evaluated. The European Economic Community has been formulating new legislation on this important topic which may well provide a lead for many other countries; the legislation has been outlined in the *GAIN Report* (USDA, 2009).

Selection of appropriate pesticides

This volume covers a wide range of crop species grown in tropical environments and because of the diversity of pests, pathogens and other disorders, specific recommendations for their control have not been made in the text. Many countries have extension or advisory officers who are aware of what is nationally or locally available and it is anticipated that appropriate advice will be increasingly available to subsistence farmers as both extension coverage and farmer training develops further.

There is an increasing demand and need for farmers (including subsistence farmers) to be more aware of the concept and practical applications of IPM.

Integrated Pest Management (IPM)

Over the years, many farmers have been using increasing amounts of natural and synthetic pesticides. The use of these materials has often been led by the field advisors of the pesticide manufacturers and to some extent by the farmers themselves, who, if they have finances available, have preferred to use a chemical to control a pest as soon as it appears rather than wait to see if it builds up and becomes a problem in the crop. This philosophy tended to prevail until towards the end of the 20th century, it largely ignored the importance and value of natural predators and also endangered hive and wild bees as well as other pollinating insects which are vital for fruit and seed production in many crop species. The agrichemical industry had provided the grower and farmer with a very large 'armoury' of pesticides for the control of the whole spectrum of pests, pathogens and weeds, and this developed more momentum during the 1940s with the availability of synthetic chemicals for agricultural and horticultural use. During the latter part of the 20th century the costs and undesirable effects of excessive pesticide use was realized by a wide range of interested parties including environmentalists, consumers and even farmers; this was especially emphasized by increased awareness of the long-term residual effects of the chlorinated hydrocarbon and organophosphate pesticides. This started to influence the attitudes of consumers, environmentalists and the farming communities towards pest control methods and regimes.

The use of pesticides can result in one or more of the following:

- Added cost of materials and labour for their application by the farmer.
- Reduction of pollinating insect populations.
- Pests and/or pathogens developing strains which are resistant to specific pesticides.
- Hazards and dangers to the persons handling and applying the materials – this has been especially significant in the tropics where there is a reluctance to wear recommended protective clothing or follow manufacturer's instructions regarding safety. In some instances the author has seen farmers using agrichemicals which do not even have clear instructions in the farmer's own language for preparation, application or appropriate target crop! Figure 6.1 illustrates some safety issues when handling pesticides. Safe use of all types of chemical pesticides is of paramount importance for the safety of the operators, consumers and the environment. Subsistence farmer training in the topics which relate to the safe use of pesticides should receive priority down to village level.
- Chemical residues on the surfaces of, or systemic within harvested crops – the export markets have tended to lead the way in reducing this hazard following increased monitoring and controls in the recipient countries, but the message has not always been emphasized to the subsistence farmer because they are not the exporters. However, this is especially important for

Fig. 6.1. Safe preparation of pesticides. Note the personal protective equipment (PPE) including gloves worn by the operative and his assistant, also the measuring vessel for accuracy of dilution rate. Photograph courtesy of The Real IPM Company (K) Ltd (www.realipm.com).

the subsistence farmer and his dependents because they tend to eat produce from the same plot day after day, year in and year out and are therefore more exposed to toxic chemical levels where there has been careless or inappropriate use which may result in significant pesticide residues on or in their produce, soils or other growth media.

One of the most significant outcomes of the points listed above has been the development of IPM. It started gathering momentum in the last quarter of the 20th century but there is an urgent need to make it widely understood and adopted by more farmers, including subsistence farmers.

Definition of IPM

'IPM may be defined as the combination and integration of approaches to pest management which maximizes real profitability and genuine sustainability for the users and farming system and gives due regard to the environment' (Whitten, 2008). However, there are additional biotechnological developments and other possibilities which some authors include in their definitions. For example some include the development of genetically modified organisms (GMOs) as cultivars with specific resistances and also the production of specific pest- and/or pathogen-controlling GMOs.

Some IPM specialists refer to specific control measures as *tactics*. These would include for example chemical, biological or cultural aspects of IPM. A single tactic or combination of tactics may be recommended depending on the pest or combination of pests to be controlled. The adoption of some tactics, such as the field insectary, or the use of biological control which targets a specific pest, can assist the survival and further multiplication of beneficial insects.

Examples of 'tactics'

Manipulation of habitat

An example of this tactic is the *field insectary* (illustrated in Fig. 6.2), where nectar-bearing flowering species are intercropped with a single crop or several crop species.

The field insectary is designed to have a range of plants that hosts the pest and that will attract natural predators or parasitoids. Therefore all three plant species shown in the example in Fig. 6.2 (maize, marigold and sunflower) were deliberately planted and they can have multiple roles in a field insectary. Maize has aphids and these in turn attract parasitoids like *Aphidius* spp. All three plants produce pollen and nectar, attracting pollen- and nectar-feeding

Fig. 6.2.   A field insectary in Kenya. Photograph courtesy of The Real IPM Company (K) Ltd (www.realipm.com).

insects such as hover flies. Field insectaries can also work as 'push-and-pull' strategies. The marigold is 'pulling' thrips out of a potential crop as they are attracted to the marigold in preference to the crop. Then the thrips in the marigold attract predators like *Orius* spp. (minute pirate bugs). Sunflowers can attract whitefly which in turn attract the parasitoid *Encarsia formosa*. However, field insectaries need sensible and observant management. If the pest levels get too high they can 'spill' into the crop. Therefore they should not be planted and ignored. This is a clear example of the necessity of farmer training for pest recognition, ability to evaluate the pest population build up and to make economic decisions (Dr H. Wainwright, Maderaka, Thika, Kenya, 2010, personal communication).

Mechanical barriers

Mechanical barriers can take the form of insect-proof nets over high value or more vulnerable crops. The nets can be over the top of a crop and supported by low frames erected in the plot before or soon after crop emergence; it is important that the net does not weigh the crop down, especially during or after rainfall. Where the multiplication of virus- and other pathogen-free or indexed plant material is taking place, insect-proof netting is used to clad a more permanent structure, such as a screen house.

Traps

Traps used under field conditions are usually homemade and consist of a yellow material covered with a sticky substance, attached to vertical posts; insects are attracted by the yellow colour. Their main purpose is to trap harmful flying insects such as whitefly, aphid, thrips, sawfly and adult leaf miner members of the Lepidoptera. However, these traps are not selective and can have the disadvantage of catching beneficial insects such as *Diglyphus isaea*, which is a leaf miner parasitoid. Smaller versions of the yellow and sticky card traps are hung in protective structures where virus vectors may have gained entry despite having insect-proof cladding. These traps are distinct from specialized traps designed to monitor insect population build up such as pheromone traps or night lamp traps.

Biological control

Biological control is the application of natural enemies to control pests, pathogens or weeds. Such measures are mostly target specific for 'pest' genera or species, and this definition includes:

- parasitoids (e.g. *E. formosa* to control whiteflies);
- bacteria (e.g. *Bacillus thuringiensis* to control Lepidopteran pests);
- fungi (e.g. entomophthoraceous fungi such as *Trichoderma viride*); and
- nematodes (e.g. *Phasmarhabditis hermaphrodita* which is used to parasitize slugs).

Chemical pesticides

These may be natural or synthetic in origin.

Agronomy

There are several agronomic methods, which may also be referred to as *cultural methods*. These include:

- tilling and hoeing to control weeds;
- crop rotation;
- grassing down for a season, or a longer period, in order to reduce the virulence of a soilborne pest or pathogen; and
- using cultivars with resistance to specific pathogens.

The need to encourage subsistence farmers to become involved in IPM

Although the concept of IPM is understood and followed by some of the larger commercial farmers in the tropics and subtropics, there is a vital and urgent need for the subsistence farmers to be encouraged to engage with it. One drawback is that many farmers are loath to see a small population of a potentially harmful organism allowed to develop at all, so would rather attempt to eliminate the pest with chemicals than to wait and see if it becomes economically important or is controlled by natural predators. Some countries have already embarked on IPM programmes with some success. National programmes targeted at subsistence farmers are needed to:

- ensure that the topics directly related to IPM are adequately covered in the syllabuses of technical, college and degree courses related to agriculture and horticulture;
- include IPM in the terms of reference and work schedules of extension and advisory officers;
- produce trainers specializing in IPM who can become resource persons in farmer training, including farmer field schools;
- encourage farmers and other workers at all levels to identify pest and pathogens in the field and comprehend their life cycles so that realistic pest management systems can be developed; and
- encourage workers at all levels to make direct and practical contact with subsistence farmers at village level.

The need for promotion of IPM

Workers at the International Crops Research Institute for the Semi-Arid Tropics (ICRISAT) surveyed farmers in India and Nepal to determine current plant protection practices by farmers at village level (Rao *et al.*, 2009). Their study found that 93% of the farmers in India and 90% in Nepal had adopted chemical control for the management of various pests in different crops; however, less than 20% of the farmers expressed confidence in their effectiveness. In India 52% of farmers obtain their plant protection advice from pesticide dealers, while in Nepal the majority of farmers (69%) make their plant protection decisions following advice from agricultural officers. The authors further found that 73% of farmers in India and 86% in Nepal initiate their plant protection activities based on the first appearance of the pest, irrespective of its population, the crop stage and their damage relationships. They also found a very low percentage of farmers using protective clothing while spraying and those farmers reported pesticide-related health problems. It was found that IPM practices in selected villages brought a 20–65% reduction in pesticide use in different crops.

Physiological Disorders

The physiological disorders of vegetables are sometimes referred to as 'non-parasitic disorders'. These disorders can result from any of several causes resulting from plant nutrition, weather, and plant (crop) management and include:

- nutrient deficiencies (macro- and micronutrients) arising from soil nutrient imbalance or soil pH affecting micronutrient availability;
- toxicities, arising from pesticides, nutrient imbalance, soil pH;
- physiological problems and blemishes relating to water relations, high temperature (including sun scald) and low temperature (including frost damage); and
- fertilizer scorch, direct spray damage, spray drift or soil salt levels.

Concept of Organic Vegetable Crop Production, its Requirements and Methodology

The reference to 'organic vegetables' is to their production method. The principle is that the consumable crops are produced without the use of specified crop protection chemicals or inorganic fertilizers on land that is defined as suitable (i.e. approved by an overseeing organization). This latter point is intended to ensure that the market label is authentic.

The development of market outlets for organic produce has created a demand for organically produced vegetables. Some governments have aimed to increase the percentage of crops produced organically which in turn is expected to increase the market needs for organically produced vegetables. As a result of widespread propaganda there may well be subsistence farmers who would like to engage in this vegetable crop system.

Although only a few subsistence farmers may start producing to satisfy their own requirements or to produce for this potential market, it is important that the concept is understood so that it can be debated at grower level. This may be especially useful if a niche market for organic produce arises, for example in the local tourist industry or local markets where consumers have been encouraged to turn to organically produced vegetables.

Some large-scale growers in the tropics and subtropics have entered into contracts to supply markets in developed countries. The International Federation of Organic Agriculture Movements (IFOAM), with its head office in Bonn, Germany, has a global role with links to member organizations in some 108 countries. Generally, it is a prerequisite that organically produced seed has to be used for the production of authentic 'organic crops'. IFOAM liaises with the main international seed organizations so that there is a sufficient supply of organically produced seed available for those growers who may choose to produce authentically produced organic vegetables, although the seed may not always be available in all areas, especially where there is no organized seed industry activity. As with all official schemes, and with consumer protection in mind, there has to be an overseeing organization or agency in each country subscribing to organic food production to monitor the authenticity of organically produced crops. In some countries during the early stages of the establishment of organic vegetable production there were derogations in place which allowed producers of vegetables marketed as 'organic' to use non-organically produced seed (i.e. 'orthodox seed').

Postharvest Deterioration and Potential Losses of Vegetable Crops

Postharvest loss may include reduction of bulk or weight, reduction of nutrient values, significant loss of overall quality which increases waste during culinary preparations and reduction of product value.

The deleterious factors which contribute to postharvest losses of vegetable crops starts in the field as soon as an individual crop is picked, or cut from the mother plant, or lifted from the soil. Generally when considering subsistence farmers we assume that all that is produced will be utilized by the farmer, his or her dependents and also any livestock maintained by the farmer. Knowing that some surplus may either be exchanged or be traded at the 'farm gate' we must also assume that the postharvest duration may be longer than simply 'plot to pot'.

The distance and duration increases when the potential sale of surplus crops in local markets becomes realistic for the more successful or enterprising farmer.

Unlike the commercial vegetable producers and marketers who can procure marketable crops from more than one location, the subsistence farmer has to rely on produce from a single location, therefore crop failures and postharvest losses can result in a significant reduction of food supply. There are several factors which may contribute to postharvest loss and more than one may occur at any one time.

Examples of reasons for postharvest loss of crops are:

- Not using the optimum cultivar for the season and potential storage – this point is of particular importance when considering distances from market and transport conditions.
- Short-term holding or longer term storage in adverse environmental conditions – for example:
    - Some produce may be predisposed to further damage as a result of physiological damage during its development, e.g. some tomato fruit

may display equatorial splitting as a result of irregular rainfall or irrigation during late fruit development, such cracks predispose the fruit to splitting and further pathogen entry. Figure 6.3 illustrates tomato fruit cracking and also shows secondary infection starting.

- Leaving harvested produce unprotected in the sun and/or dry wind on the headland. Cut or picked commodities should be immediately placed into shaded containers such as plastic buckets or boxes, preferably lined with some soft material to minimize bruising. Only root and tuber crops should be allowed to come into contact with the soil during the immediate postharvest period.
- Tuber and root crops breaking dormancy during storage.
- Harvesting produce while wet from rain or irrigation.
- Mechanical damage during lifting, picking or handling – this includes:
  - Damage by forks, spades or other implements; mechanical damage predisposes the produce to postharvest pests and pathogens as well as devaluing the produce.
  - Unsatisfactory handling which causes bruising (often not identified until after storage life has commenced).
- Pests and pathogens – these may already be apparent either before or during harvest, for example *Botrytis* or *Peronospora* on some leaf crops such as lettuce and cabbage.
- Infection from saprophytic storage pathogens.
- Damage resulting from storage pests including rodents.

Fig. 6.3. Split tomato fruit, showing postharvest disease developing in the crack.

Outlets for Surplus Produce and Start of Market Development

Sales of surplus produce include: farm-gate sales, farmer markets, roadside markets, purchase by collection traders, assembly markets, wholesale markets and cluster farming. Some of these are discussed in more detail below.

Farmer markets

These are very traditional markets, usually established in urban areas. In many cases there are simple stands provided for sellers' use. The main advantage to farmers is that they are able to make sales direct with the consumer; although having the produce displayed and exposed can lead to a fall in value if the produce is not sold before the heat of the day takes its toll on quality. Customers handling, or mishandling, the produce may well accelerate the product's deterioration.

Roadside markets

These are usually impromptu: local farmers or members of their household take advantage of passing traffic, either vehicular or on foot, who are attracted by the display and apparent freshness of the produce.

Purchase by collection traders

In this type of outlet the farmer sells to the trader either on the farm site or at a local market. In many of these scenarios the trader is also a moneylender and is generally in a position to have his own transport, which when considered together can result in low returns for the farmers who may have been obliged to opt for this system.

Assembly markets

This type of market is where the grower sells his or her produce direct to a wholesale trader. The advantages to the farmer are that relatively large quantities of surplus produce can be sold. The main advantage to the trader is that the growers bring their produce to the market thus saving him travelling around the countryside seeking out available produce. Farmers' cooperatives may see assembly markets as an initial breakthrough beyond the farmer markets.

Wholesale markets

This type of market can be seen as an assembly point for produce already procured by traders, or delivered to the market site by, or on behalf of farmers or their cooperatives. Purchasers of the produce can be wide ranging in their retail interests, for example hawkers, small stallholders, private purchasers or small shops. Some of the larger wholesale markets have the capability of sales and distribution of produce and products from further afield; these may include fruit, nuts and agricultural or horticultural commodities needed by urban populations and rural communities. In some markets, seed companies or government agricultural departments may have vegetable and other crop seeds for sale.

Cluster farming

It has been reported by Montiflor (2008) that in the case of smallholder farmers in Southern and Northern Minanao in the Philippines *cluster farming* has become an important system by which farmers supply their produce to institutional consumers such as hotels, hospitals, supermarkets, restaurants and fast food chains. This is a move significantly advanced from individual farmers finding a niche market related to a seasonal or local tourist industry. The vegetable farmers in the Philippines generally cultivate less than 3 ha (but this is considered to be significantly more than the majority of subsistence farmers in many other parts of the world). The individual farmer who joins the *cluster* is neither able to meet the quantity or constant quality which these potential sales outlets require, nor are the farmers individually able to compete with produce imported from other areas or abroad. To achieve the necessary capability farmers have formed groups or so-called 'clusters'. These could possibly be described loosely as a form of cooperative without all the official organizational requirements.

It has been reported that the liaison between farmers and the institutional outlets has resulted in farmers coordinating their cropping programmes, sowing or planting and even coordinating the adoption of appropriate cultivars. This level of cooperation and coordination achieves a greater volume or bulk of the various institutions' specific requirements.

The socio-economic impact that the establishment of the clusters has achieved is interesting. Montiflor *et al.* (2009) found the following non-monetary benefits:

- access to the larger institutional markets;
- improved quality of market information;
- links between the markets and the farmers' production;
- improved technical and financial support; and
- improved access to crop production inputs.

The above achievements cited by Montiflor *et al.* (2009) can also be regarded as some of the advantages of any successful marketing system and serve to emphasize the importance of a well-organized marketing scheme to improve the lot of small farmers and their dependents.

Development of Small Farmer's Marketing Opportunities

Global and national awareness of the high nutritional value of horticultural crops in the human diet have been largely responsible for the urban interest in vegetables and other horticultural crops such as fruit. This, along with increased urban populations, has led to the increased possibilities for small growers to produce crops for market in addition to their own dependents' essential subsistence. The small farmer is able to start to become more able to support schooling, healthcare and other necessary requirements of his or her dependents and to rise above poverty when the marketing of his or her crops becomes viable. Thus, the development of efficient marketing processes is vital for the commercial success of the small grower; this is especially important in the tropics where deterioration of produce can be rapid.

The formation of farmers' cooperatives has often been described as a positive way forward for farmers with relatively small land areas under cultivation. However, as Sargent (1975) found after investigating why some vegetable cooperatives had not been a success, their initiation should only occur on the basis of a satisfactory feasibility study, and even then, their success will depend on the training, experience, initiative and honesty of the management, and the number and loyalty of members.

Further Reading

CABI (2010) *CABI Compendia (Beta)*. Compendia of interactive encyclopedias. Available at: www.cabicompendium.org/cpc (accessed 16 June 2010).

Cull, D.C. and George, R.A.T. (1981) Vegetable production on small farms overseas. In: Spedding, C.R.W. (ed.) *Vegetable Productivity*. Macmillan, London, pp. 186–189.

Food and Agriculture Organization (FAO) (1989) *Prevention of Post-harvest Food Losses: Fruits, Vegetables and Root Crops*. FAO, Rome, Italy.

International Society for Horticultural Science (ISHS) (2008) *Proceedings of the II International Symposium on the Socio-Economic Impact of Modern Vegetable Production Technology in Tropical Asia. Acta Horticulturae 794*. ISHS, Leuvin, Belgium.

International Society for Horticultural Science (ISHS) (2009) *Proceedings of the International Symposium on Improving the Performance of Supply Chains in the Transitional Economies. Acta Horticulturae 809*. ISHS, Leuvin, Belgium.

Mathur, S.B. and Kongsdal, O. (2003) *Common Laboratory Seed Health Testing Methods for Detecting Fungi*. International Seed Testing Association (ISTA), Basserdorf, Switzerland.

Montiflor, M.O. (2008) Cluster farming as a vegetable marketing strategy: the case of smallholder farmers in Southern and Northern Minanao. *Acta Horticulturae* 794, 229–238.

Montiflor, M.O., Batl, P.J. and Murray-Prior, R. (2009) Social impact of cluster farming for smallholder farmers in southern Philippines. *Acta Horticulturae* 809, 193–200.

Proctor, F.J., Goodliffe, J.P. and Coursey, D.G. (1981) Post-harvest losses of vegetables and their control in the tropics. In: Spedding, C.R.W. (ed.) *Vegetable Productivity*. Macmillan, London, pp. 139–172.

Richardson, M.J. (1990) *An Annotated List of Seed-borne Diseases*, 4th edn. International Seed Testing Association (ISTA), Basserdorf, Switzerland.

United States Department of Agriculture (USDA) (2009) *USDA Foreign Agricultural Service GAIN Report No. E 49002*. USDA, Washington, DC. Available at: http://www.useu.usmission.gov/agri/pesticides.html (accessed 16 June 2010).

7 *Alliaceae* – Onions and Related Crops

The important vegetable crops in this family which are cultivated in the tropics include:

- *Allium cepa* L. – onion and shallot;
- *Allium sativum* L. – garlic;
- *Allium fistulosum* L. – Japanese bunching onion, Welsh onion;
- *Allium tuberosum* Rottl. ex Spreng – Chinese chives; and
- *Allium chinense* G. Don. syn. *A. bakeri* Regel – rakkyo.

The two most important are onion and garlic. Shallots (formerly considered as *Allium ascalonicum* L. were earlier regarded as a separate species from onion but are now considered by taxonomists to be *A. cepa*). Japanese bunching onion is produced from seed or vegetative division. Garlic and rakkyo are traditionally vegetatively propagated by bulbs ('cloves'). Chinese chives are generally multiplied by bulbils formed as a result of apomixis.

The physiology of onion and garlic has been reviewed by Brewster (1997) and the scientific principles as related to the practice of the production of onions and other vegetable alliums, their biochemistry and food science have been described and discussed by Brewster (1994).

The medicinal values of the vegetable alliums in the diet have been recognized and have subsequently increased their popularity (Block, 1985). The bulbs of onion, shallot and garlic can be stored and can therefore significantly contribute to food security.

Allium cepa (Onion)

Allium cepa probably originates in Afghanistan, Iran and Pakistan but is now widely cultivated in the tropics. The species is a biennial but it is cultivated for its edible bulb formed at the end of the first year. The storage potential of the mature bulbs has led to onions becoming an important culinary crop. The very young plants, developing and mature bulbs are each used in a range of culinary ways.

Cultivars and types

As onion bulb formation is dependent on day length, there are specific day-length requirement groups for different latitudes, from types with a 16h day-length requirement adapted to the northern and southern latitudes to 12h day-length types suited to the tropics. It is important that cultivars introduced to the tropics are of the appropriate bulbing response group for the latitude where it is proposed to grow them. Currah and Astley (1996) have discussed possible ways that onion cultivars and germplasm descriptors for day length can be improved. Temperature also has an important influence in that the rate of bulb formation increases with temperature increase. Onion physiology, growth, development and bulbing have been reviewed by Brewster (1994).

Soil, drainage, pH and nutrient requirements

The sites used for onions and related crops should not have produced a related crop in the recent rotation.

These crops can be produced on a range of soil types, ideally a soil with good potential moisture retention is preferable because onions have a relatively shallow root system and require fairly continuous available water until completion of bulb formation, but the crop cannot tolerate wet soil conditions; if these are likely then the crop should be produced on raised beds.

Onion crops respond well to bulky organic manures incorporated during site preparation. However, caution should be taken if liming is also required during preparation by not applying the lime and organic materials together. The onion crop's optimum soil pH range is 6.0–6.8.

The onion crop responds to N:P:K base dressings of 1:2:2 applied at a rate of $40\,g/m^2$ during seedbed preparation although the nutrient levels of any bulky organic manures which have been included should be taken into account and the inorganic fertilizer reduced accordingly.

Sowing and crop establishment

The onion crop is mainly produced from seed although some growers produce it from 'sets'. Onion seed has a relatively short storage life especially in open storage in the tropics, it is therefore advisable, as far as is possible, to use vapour-proof packaged seed previously unopened which has been obtained from a reliable supplier.

The seedbed should be firm but with a fine tilth. Onion seedlings are poor competitors with weeds, because of their initial single cotyledon emergence.

Ideally the stale seedbed technique should be used prior to seed sowing so as to minimize early competition between onion and weed seedlings. Seed is sown approximately 1.0 cm deep in drills 30 cm apart; the plants are subsequently thinned within the rows to approximately 10 cm apart. However, plant population per unit area significantly influences final bulb size (assuming timely and efficient weed control) and some growers prefer closer spacing within the rows to increase total onion bulb yield by weight.

Harvesting

Bulbing onions mature in approximately 3–5 months from sowing; the timing to bulb maturity depends on the cultivar and local climate. Those grown as bunching ('spring onions') can usually be harvested 5–7 weeks from sowing. Any plants which flower during the first year should not be retained for seed production. When the bulbs get near to maturity the leaves tend to collapse at the necks and from this stage onwards the foliage starts to die back and the mature bulbs commence their ripening process. Any supplementary irrigation should cease by this stage. When approximately half the plants have weakened at their necks and the leaves have toppled over the bulbs should be lifted. Onion bulbs have a relatively shallow root system and the maturing bulbs can either be forked out or the roots undercut with an onion hoe or a long-handled draw hoe. Whatever method is used care must be taken not to cut or otherwise damage the bulb. In dry climatic conditions the bulbs can remain on the surface to dry and further ripen. In the event of wet conditions the bulbs should be transferred to a covered, airy shelter and exposed to the sun whenever the weather allows. As the bulbs ripen and dry the dried foliage should be removed by twisting and pulling; cutting off the tops should be well above the bulb's shoulder if removed with a knife. Care must be taken not to remove outer scales which are not dry and loose. Bulbs usually require approximately 2 weeks of ripening, ideally in the sun to be cured prior to storage.

Storage

Only bulbs which are sound and with narrow necks should be stored, the discards should be used first. Figure 7.1 illustrates the range of graded bulb qualities prior to storage. Cured bulbs should be stored under cover with good aeration on racks of mesh or slats. Several types of structure are used depending on location. In some countries the bulbs are sliced transversely and sun dried for long-term storage.

Pathogens

The main seedborne pathogens of *Allium* species are listed in Table 7.1.

Fig. 7.1. The range of onion bulb qualities found when sorting cured bulbs: bottom two rows satisfactory, centre mechanically damaged, top left group 'bull-necks' and top right 'splits'.

Table 7.1. The main pests and pathogens of *Allium* species.

Pest or pathogen	Common names
Ditylenchus dipsaci	Bloat, eelworm rot, stem and bulb nematode
Alternaria porri	Purple blotch
Botrytis allii	Damping off, grey mould, neck rot
Botrytis byssoidea	Seedling damping off, neck rot
Botrytis cinerea	Grey mould, collar rot, leaf spot
Cladosporium allii	Leaf blotch
Colletotrichum circinans	Smudge, damping off, anthracnose
Fusarium spp.	Fusarium basal rot
Peronospora destructor	Downy mildew
Pleospora herbarum	Black stalk rot, leaf mould
Puccinia allii	Rust
Sclerotium cepivorum	White rot
Urocystis cepulae	Smut
Onion mosaic virus	
Onion yellow dwarf virus	

Allium cepa var. *ascalonicum* (Multiplier Onion Group; Shallot)

Although now said by taxonomists to be a type of *A. cepa*, shallots are believed to have developed from *A. cepa* as a form which has become vegetatively propagated. Growers continue to regard shallots separately from onions because they are generally vegetatively propagated from the bulbs produced by natural division. Shallots can be produced from seed as well as bulbs outside the tropics but are mainly produced from bulbs in the tropics. Some tropical areas, for example Indonesia, have found it easier to produce a satisfactory shallot crop than onions; shallots also have a shorter growing period and there is also a culinary preference for shallots in some communities. Research work has continued to produce the shallot from true seed, and some success has been reported from Ethiopia, Sri Lanka and Indonesia (Currah and Proctor, 1990).

The shallot is a perennial but grown as an annual from bulbs. The mother bulb produces a cluster of approximately five to eight off-sets. The crop is generally not successful in forest areas or sites with frequent rainfall.

Soil

A relatively light sandy loam with good drainage is preferable. Optimum pH is between 5.6 and 6.5. Where soils are likely to be poorly drained the crop is produced on raised beds. Growers incorporate composted materials during site preparation followed by firming the planting area. A suitable base fertilizer dressing is N:P:K 1:1:1 applied at the rate of $20\,g/m^2$.

The daughter bulbs are usually pressed into the soil in rows 30 cm apart and 12 cm within the rows. There can be a tendency for the emerging roots to force the planted bulb up and out of position; this can usually be avoided by planting the bulbs with a trowel up to the base of their necks and firming in, although this takes more time. Weed control is especially important during the early life of the crop as the linear foliage does not provide significant competition to young weeds.

Harvesting

The crop is usually mature after approximately 12 weeks from planting although this can vary with cultivar and location. Maturing clusters of bulbs show signs of yellowing or browning from the leaf tips downwards. Irrigation should cease by this stage. After several days of foliage senescence the crop should be lifted and left to dry *in situ* as illustrated in Fig. 7.2, but in rainy conditions the crop should be moved to an aerated shelter for further drying and curing.

Storage

Bulbs are usually tied in bundles and hung off the ground in a ventilated but rainproof shed.

Pests and pathogens

These are as listed for *Allium* in Table 7.1.

Fig. 7.2.   Harvested shallots lifted, bunched and drying off and curing in the field.

Allium sativum (Garlic)

The 'mother' garlic plant is a composite bulb consisting of many smaller wedge-shaped 'daughter' bulbs each of which may be referred to as a 'clove'. The single clove is the propagule used to produce a single plant in the following crop or generation.

There are two main types of garlic: (i) the hard-neck; and (ii) the soft-neck. There are many clonal selections of each based on number of cloves, overall bulb size and pigmentation.

Quality of planting material

There has been increased interest in recent years in the provision of pest- and pathogen-free garlic planting material for small farmers (FAO, 2010). In the case of garlic the main pathogens which can be transmitted via planting material are:

- onion yellow dwarf virus (OYDV);
- garlic yellow streak virus (GYDV); and
- leek yellow stripe virus (LYSV).

Mosaic in garlic, which shows as leaf chlorosis, especially in young leaves, can be transmitted from plant to plant by aphids or via infected mother bulbs in the planting material. Individual plants showing symptoms should be removed from the field crop, especially if intended for further multiplication for retention as planting material.

Field requirements

As with other vegetable alliums, crop rotation should be followed closely as there are several economically important soilborne pests and pathogens.

Garlic grows best in cool soils and this is usually achieved by mulching after planting; temperatures up to 30°C are suitable when the cloves have become established.

Planting patterns and densities depend on local irrigation methods and weed control. It is common practice to grow the crop in beds the centres of which can be conveniently reached from either side; this is particularly important for efficient hand weeding.

The cloves are planted vertically, in firmed beds, so that the top of the clove is just below the soil surface, at 10–15 cm within the rows and 30 cm between the rows.

The crop should be kept weed free, especially in its early stages as the leaves do not form a canopy to suppress indigenous weeds.

Harvesting and storage

When the crop is approaching maturity, usually 3–4 months from planting, the leaves start to die back from their leaf tips, the neck weakens at the base and topples over. Some growers accelerate the drying-off process by pushing the foliage over. Following lifting, the plants are dried under cover on raised nets or mesh in an airy and rain-free shelter. Sun drying is ideal, although care should be taken to ensure that they are not left out in rainy conditions.

When the crop has dried off, the bulbs are cleaned by hand and any remaining dry foliage and stems removed.

The cleaned bulbs can be stored at approximately 25°C and relative humidity of less than 70% for about 3 months.

Allium fistulosum (Japanese Bunching Onion, Stem Onion, Welsh Onion)

This *Allium* species originates from China and is an important crop widely grown in Asia. It is cross compatible with *A. cepa*. Plant breeders have produced lines which have increased winter hardiness for production in colder areas of countries including China, Nepal and North Korea where despite very high temperatures in their summer seasons the winters can be severe. *Allium fistulosum* which is used as a 'bunched onion' is distinct from the so called 'spring onions' or 'salad onions' which are younger plants of *A. cepa*. Basal buds produce a cluster of new plants. There are tall and short leaf types, with clonal differences within these two types. The common name 'Welsh onion' is derived from the Anglo-Saxon *welise* and the German *welsche* meaning foreign; the species does not originate in Wales, UK.

Environmental requirements

Allium fistulosum is more successful as a crop at higher altitudes of 1000 m and above than the lowland areas and can tolerate wetter conditions than *A. cepa*, but not waterlogging.

Soil

The site and soil requirements and also preparations are as described above for *A. cepa*.

Crop establishment

The crop produces tillers and can be propagated by seed or division. Division is preferred by most growers in order to ensure that they have the preferred clone, especially where winter hardiness is an important requirement; some forms do not readily produce true seed. Single plants are planted in a square spacing 20 × 20 cm or 25 × 25 cm depending on the vigour of the clone. When planting material is produced from seed the seedlings are planted in groups of three or four approximately 20 cm each way.

Harvesting

The crop can be harvested from 3–4 months or otherwise left in the soil for longer, but preferably it should be harvested or lifted for further division when there is a significant amount of leaf senescence.

Subsistence farmers can harvest as required and remove individual plantlets by pulling them out, sideways, from the main clump. Harvesting can be coupled with replanting separated plantlets at intervals to ensure continuity of supply. The crop is not normally stored.

Pests and pathogens

The pests and pathogens of *A. fistulosum* include those listed in Table 7.1. The following pests are also important:

- *Spodoptera exigua* – lesser cotton leaf worm;
- *Spodoptera littoralis* – cotton leaf worm; and
- *Thrips tabaci* – thrips.

Allium tuberosum (Chinese Chives)

This species originates in Eastern Asia where it is still found in the wild but it is now widely cultivated in much of Asia. Two types are recognized, the large leaf type 'Big leaf' and the smaller leaf type. The large leaf type is very popular in China and Taiwan especially when grown as a blanched crop.

Plant production and planting

Transplants are either raised from seed or by division of the clump which forms tuberous rhizomes; these are separated into individual plantlets. The same spacings as described above for *A. fistulosum* are generally adopted.

Harvesting

The young leaves and flowers are used for culinary purposes although flowering can be sparse in tropical lowlands where seeds do not always develop. It is usual for subsistence farmers to harvest the leaves as required.

Pests and pathogens

These are listed in Table 7.1.

Allium chinense syn. *A. bakeri* (Rakkyo)

Rakkyo originates in Central and Eastern China and it has become popular as a cultivated crop and is widely grown in India, Taiwan and China and is an important commercial crop in Japan. There are some named cultivars, some of which have been developed in India and Taiwan. Rakkyo forms clumps of daughter (lateral) bulbs which are small and ovoid. The bulbs are widely used for the production of pickles and savoury preserves.

Crop production

Rakkyo succeeds on light soils but will benefit from well-decomposed compost which has been incorporated during site preparation. The majority of growers produce their crop from the mature bulbs although some smallholders in Southern China prefer to grow the crop from seed which is sown in midwinter, harvesting the new crop the following spring (Herklots, 1972). The spacing is the same as stated below for crops produced from bulbs although plant density within the rows can be increased. When produced from bulbs the propagules are planted during late summer 8 cm apart, in rows approximately 60 cm apart. The crop from this planting is usually harvested in the early summer of the following year; although some growers leave the plants in the ground until their second year before they are lifted for harvest.

Pests and pathogens

These are listed in Table 7.1.

Further Reading

Block, E. (1985) The chemistry of onions and garlic. *Scientific American* 252, 94–99.

Brewster, J.L. (1994) *Onions and Other Vegetable Alliums*. CABI, Wallingford, Oxon, UK, 236 pp.

Brewster, J.L. (1997) Onions and garlic. In: Wien, H.C. (ed.) *The Physiology of Vegetable Crops*. CABI, Wallingford, Oxon, UK, pp. 581–619.

Currah, L. and Astley, D. (1996) Describing onions: can the definition of day-length descriptors be improved? In: Currah, L. and Orchard, J. (eds) *Onion Newsletter for the Tropics 7*. Natural Resources Institute, London, pp. 77–81.

Currah, L. and Proctor, F.J. (1990) *Onions in Tropical Regions*. Natural Resources Institute Bulletin No. 35. Natural Resources Institute, Chatham Maritime, Kent, UK.

Food and Agriculture Organization (FAO) (2010) *Quality Declared Seed for Vegetatively Propagated Crops*. FAO Plant Production and Protection Paper. FAO, Rome, Italy.

Walkey, D.A.G., Webb, M.J.W., Bolland, C.J. and Miller, A. (1987) Production of virus free garlic (*Allium sativum* L.) and shallot (*A. ascalonicum* L.) by meristem-tip culture. *Journal of Horticulture Science* 62, 211–220.

Cruciferae – Crucifers

This family contains important vegetables including:

- *Brassica oleracea* L. – the brassicas or cole crops;
- *B. oleracea* var. *capitata* L. – cabbage;
- *B. oleracea* var. *botrytis* L. – cauliflower;
- *Brassica campestris* L. – turnip and related crops;
- *B. campestris* subsp. *rapifera* Metz. – true turnip;
- *B. campestris* subsp. *chinensis* Jusl. – Chinese mustard, pak-choi;
- *B. campestris* subsp. *pekinensis* (Lour.) Rupr. – Chinese cabbage, pe-tsai;
- *Brassica juncea* (L.) Czern. and Coss. – Indian or brown mustard, leaf mustard;
- *Brassica nigra* (L.) Koch – black mustard; and
- *Sinapis alba* L. – white mustard.

These 'mustards' are generally cultivated for the production of condiments which are obtained from the crops' seeds, although *B. juncea* is also cultivated as a leafy vegetable in addition to being an important oilseed and condiment mustard.

Eruca vesicaria (L.) Cav. subsp. *sativa* (Mill.) Thell. (tamarina, rocket salad) is a species that is cultivated in India and some other parts of Asia as an oilseed crop; it is frequently cultivated as a green vegetable and stock feed.

There are in addition to the above vegetable crops, some economically important agricultural crops including oilseed and fodder species.

The cultivated radish *Raphanus sativus* L. is another important vegetable in this family.

Cole Crops

This group of *B. oleracea* types is generally considered to include cauliflowers (including the so-called broccoli), kales and cabbage (but not *B. campestris* subspecies, which are the Chinese cabbage types). The general history of the main types has been discussed by Thompson (1976).

The most important cole crop cultivated in the tropics is *B. oleracea* var. *capitata* widely referred to as cabbage or white cabbage. The other types of *B. oleracea* which are prominent in temperate vegetable growing schemes can often be grown successfully at higher altitudes in the tropics but are not considered here as crops normally important for subsistence farmers.

Origin and types

The cabbage has originally been developed from selections of the wild cabbage which is a native of south-west Europe and the lands around the Mediterranean. The morphological range of cabbages and their suitability of individual cultivars for different seasons and environments is a fine example of cultivar development by selection and subsequent plant breeding. There is a range of head shapes, red pigmentation and degrees of leaf texture.

Although this species originates as a temperate vegetable it is becoming increasingly popular as a useful tropical crop especially in the upland regions. It is widely grown in parts of South America, Asia, including the Philippines, Indonesia, the Indian subcontinent and also parts of Africa.

Asian plant breeders, especially in Japan, have improved tropical cultivars with increased resistance to some pathogens such as black rot (*Xanthomonas campestris* pv. *campestris*), and have also developed hybrid cultivars. However, from a subsistence farmer's point of view, uniformity of maturity is not always a desirable feature unless grown as a cooperative crop or there is a local market where any surplus can be sold or traded with economic advantage to the producer.

Crop establishment

This crop requires a soil with a pH of 6.0–6.5; any required liming material should be incorporated during site preparation. Acid soils increase the incidence of club root (*Plasmodiophora brassicae*) and reduce the availability of boron and molybdenum (both essential micronutrients).

Transplants are usually produced in beds for transfer to their final quarters when large enough to handle, usually 4–5 weeks from sowing. Seed is sown thinly either in drills of 1 cm depth or broadcast thinly and covered with prepared fine soil, the beds are lightly firmed after seed sowing and should be kept moist if dry conditions prevail. The optimum spacing for the seedlings is approximately 5 × 5 cm.

The young plants are transplanted in their final quarters in rows 45 cm apart and the same distance apart within the rows. The spacing can be increased up to 60 cm square for the larger headed cultivars. It is possible to achieve a continuity of supply by making small sowings at approximately 5 week intervals. In some areas continuity can also be achieved by a succession of suitable cultivars to take the successive seasons into account.

Early weed control by hoeing is important to reduce competition, followed by further hoeing until the crop starts to form a leaf canopy unless mulching has suppressed the emerging weed species. Care should be taken to avoid disturbance or damage to the roots which tend to be near the soil surface.

Cabbage responds to adequate available soil water therefore, ideally, irrigation should be applied when indicated by soil moisture levels.

Harvesting

The duration from planting out to maturity largely depends on the cultivar; those known to be suitable for tropical production can be ready approximately 2 months from planting out. Open-pollinated cultivars will usually have a harvesting duration of up to a further 3 weeks; however hybrid cultivars will have a much narrower maturity range. Mature cabbage heads can be expected to hold in the field for several days but will tend to split if left too long before cutting. The individual heads are cut so that a whorl of protective non-heart leaves can protect the head postharvest.

The harvested cabbage heads can normally be stored in cool and shady conditions for a few days, but not for longer periods without low temperature storage.

Pests and pathogens

The pests and pathogens for *Brassica* species are listed in Table 8.1.

The Chinese Cabbages

Origins and types

This group originated in the Regions of China and Japan from whence it has become an important vegetable crop in Asia; the different types have also become important in Africa and South America. The botany and breeding of Chinese cabbage for the tropics and subtropics has been described by Opeña *et al.* (1988). There has been some interest in producing them in some temperate areas.

There is a wealth of cultivars and local selections available in Asia which are loosely referred to as Chinese cabbages. The material can generally be divided into one or other of the following subspecies of *B. campestris* and some cultivars show morphological characters which are intermediate between these two subspecies. There are open-pollinated and F_1 hybrids of both types.

- *B. campestris* subsp. *chinensis* (Chinese mustard, pak-choi) – types from this subspecies cultivated as vegetables include flowering and

Table 8.1. The main pests and pathogens of *Brassica* species.

Pest or pathogen	Common names
	Snails
	Rodents
Meloidogyne spp.	Root knot eelworm
Pieris canidia	Cabbage white butterfly
Plutella xylostella	Diamond-back moth
Hellula undalis	Cabbage webworm
Albugo candida	White blister
Brevicoryne brassicae	Cabbage aphid
Alternaria brassicae	Grey leaf spot
Alternaria brassicicola	Black spot, wire stem
Leptosphaeria maculans	Dry rot, blackleg, black rot
Peronospora parasitica	Downy mildew
Mycosphaerella brassicicola	Black ring spot
Plasmodiophora brassicae	Club root
Rhizoctonia solani	Wire stem, damping off
Sclerotinia sclerotiorum	Watery soft rot, drop, white blight, Sclerotinia rot
Pseudomonas syringae pv. *maculicola*	Bacterial leaf spot
Xanthomonas campestris pv. *campestris*	Black rot
Turnip yellow mosaic virus	

vegetative shoots, open heads or shoots and many of the types tend to form a rosette.

- *B. campestris* subsp. *pekinensis* (Chinese cabbage, pe-tsai) – the main vegetables in this group are the Chinese cabbage types which have a wide range of more solid heads; different head shapes are displayed by distinct cultivars and they range from almost spherical to ovate.

Soil requirements

A soil pH from 6.0–7.5 is suitable; if the pH level is lower than 6.5 liming materials should be incorporated during site preparation. One or two top dressings of suitable nitrogenous fertilizers can be applied at the rate of 30 g/m^2 following young plant establishment if significant leaching has occurred or plant growth is slow. The second top dressing should be applied approximately 10 days later at the same rate.

Crop establishment

Seed of both subspecies may be sown in seedbeds or direct into the final quarters. There are usually local preferences as to which of these methods to adopt, this depends on pressure for land, length of growing season and seedling vulnerability from the prevailing weather.

Young transplants of both subspecies are planted out, or thinned, to 30 cm apart in rows up to 40 cm apart, although higher plant densities can be used for the more upright and smaller cultivars of Chinese mustard. Some growers place a mulch over the seedbed to suppress weeds and reduce surface water loss. If either type is planted at a higher plant density there is an increased risk of crop loss due to the pathogens which are more prevalent in the humid tropics.

The developing crop should be maintained weed free; hoeing can be efficient, although further mulching with rice straw is usually believed to be labour saving.

Harvesting

The time from planting to harvesting depends on cultivar and environment. Generally the duration to harvest is for Chinese mustard 6–8 weeks and for Chinese cabbage 7–10 weeks. Figure 8.1 illustrates the bunching of Chinese mustard (pak-choi) immediately after harvesting.

Both types can be stored for a few days provided that damaged or diseased leaves are trimmed off and the produce placed in a cool shady shed or room immediately after harvesting. One of the most important global pests of *Brassica* species is *Plutella xylostella*, the diamond-back moth, as illustrated in Fig. 8.2.

Fig. 8.1.   Postharvest bunching of Chinese mustard (pak-choi) immediately following harvest.

(a)   (b)

Fig. 8.2.   (a) Adult diamond-back moth. The adult has a wingspan of 13–15 mm and a body length of 6 mm. (b) Larvae of diamond-back moth. The first instar is 1.7 mm long growing to approximately 11.2 mm by the fourth (final) instar. Photographs courtesy of The Real IPM Company (K) Ltd (www.realipm.com).

Brassica juncea (Indian or Brown Mustard, Leaf Mustard)

This leafy vegetable is very suitable for subsistence and home growers as a small number of plants will provide a suitable succession of a green vegetable crop.

Origin and types

According to Purseglove (1974), the species originated in Africa from where it was taken to Asia. It is now widely cultivated especially in China and the Indian subcontinent as an industrial seed crop and as a small-scale but important leafy vegetable. There are red pigmented types available.

Crop production

For small-scale production by a subsistence farmer, seed is usually sown in a seedbed and the young plants transferred to their final quarters as described above for *B. oleracea* var. *capitata* and cultivated in the same way.

Harvesting

Harvesting usually starts 2 months from planting, the younger leaves are taken from the mother plant, two to three leaves per harvest. Depending on local climatic conditions and selection, or cultivar grown, the plants will continue to produce leaves for up to 3 months. Pursglove (1974) states that because of the presence of the glucoside, sinigrin, the leaves should be boiled twice.

Raphanus sativus L. (Radish)

Origin and types

There are several types of cultivated radish as listed below. The first four are as described by Banga (1976):

- *R. sativus* L. var. *radicula* – this is the widely cultivated salad type, it is probably more important in the temperate areas than the tropics.
- *R. sativus* L. var. *niger* – this is the large-rooted type which is popular in Germany and also important in parts of Asia.
- *R. sativus* L. var. *mougri* – this is cultivated in Asia especially for its edible seed 'pods' and foliage used as a salad or pot herb.
- *R. sativus* L. var. *olifera* – this is the fodder radish mainly grown for stock feed in temperate regions.
- *R. sativus* L. var. *longipinnatus* (Chinese radish, mooli) – this species is an important vegetable in Asia, parts of East Africa and the Caribbean.
- *Raphanus caudatus* L. (rat-tailed radish, pod radish) – this species of *Raphanus* is popular in India and parts of Asia and is cultivated for its long seed pods which can be up to 60 cm long. Herklots (1972) refers to it as *R. sativus* var. *caudatus* Alef.

The radish types described here in this chapter are: (i) *R. sativus* var. *radicula*; and (ii) *R. sativus* var. *longipinnatus*.

Soils

Radish crops tolerate slightly acid conditions and a soil pH between 5.5 and 6.8 is suitable. The ideal soil type is sandy or silty, with a high organic content. A general compound fertilizer can be applied during site preparation at a rate of 30 g/m^2 but excessive nitrogen should be avoided as leaves will be produced at the expense of the radish 'root' (the radish 'root' is a swollen hypocotyl).

Crop production

For *R. sativus* var. *radicula* types the seed is generally broadcast over a prepared bed and raked in, followed by a light watering. A light mulch of a material such as rice straw can be applied to reduce weed emergence and to reduce surface water loss. Germination is normally 4–6 days from sowing and the seedling stand can normally be observed after about 5 days.

Table 8.2. The main pests and pathogens of *Raphanus* species.

Pest or pathogen	Common names
Alternaria alternata	
Alternaria brassicae	Grey leaf spot
Alternaria brassicicola	Black leaf spot
Alternaria raphani syn. *A. matthiolae*	Leaf spot
Colletotrichum higginsianum	Anthracnose, leaf spot
Gibberella avenacea	Root and stem rot
Leptosphaeria maculans	Black leg
Rhizoctonia solani	Damping off, canker
Xanthomonas campestris pv. *raphani*	Bacterial spot
Radish yellow edge virus	
Tobacco streak virus	
Turnip mosaic virus	

Harvesting

This radish type is harvested as the swollen hypocotyls reach maturity and this can occur approximately 1 month from sowing. The foliage of pinnate leaves is also used as a salad or pot herb.

Pests and pathogens

The main pests and pathogens of *Raphanus* species are given in Table 8.2.

Raphanus sativus var. *longipinnatus* (Chinese Radish, Mooli)

Origins and types

This variety of *R. sativus* is thought to have originated in Southern China from where it has spread to Japan and other parts of Asia. It is now an important vegetable in Asia, parts of East Africa and the Caribbean. Japanese and other plant breeders have made considerable progress with cultivar development (Shinohara, 1984). There are red- and white-rooted types with improved root shape and quality. Root quality is maintained by selection or roguing during seed production; examples of typical good and poor 'roots' are illustrated in Fig. 8.3. There are both open-pollinated and F_1 hybrid cultivars and also separate white-rooted and red-rooted cultivars.

Soil and site

A loam or silt soil with good water retention characters and a pH between 5.5 and 6.8 is ideal.

Fig. 8.3. Satisfactory Mooli 'root' (on the left) and poor quality (on the right).

Crop establishment

Seed is sown in rows 30–45 cm apart, the young plants are thinned to approximately 20–25 cm apart within the rows depending on the vigour of the cultivar. A mulch of rice straw or similar material is often applied after the seed has been sown; this retains moisture and helps to suppress weed development. Adequate water should be available during the crop's production as required when rainfall is not sufficient.

Weeds should be controlled during crop establishment but care must be taken during hoeing or other cultivations to avoid mechanical damage to the developing radish 'root' crowns.

Harvesting

The leaves are used as a salad or pot herb in some areas, although excessive picking of leaves from individual plants will lead to a reduction in their root size and weight.

Duration to satisfactory root maturity depends on the cultivar and the maturity stage and development required.

Pests and pathogens

The main pests and pathogens of *Raphanus* species are given in Table 8.2.

Further Reading

Cruciferae, general

Banga, O. (1976) Radish. In: Simmonds, N.W. (ed.) *Evolution of Crop Plants*. Longman, London, pp. 60–62.

Dixon, G.R. (2006), *Vegetable Brassicas and Related Crucifers*. CAB International, Wallingford, Oxon, UK.

Hemingway, L.S. (1976) Mustards. In: Simmonds, N.W. (ed.) *Evolution of Crop Plants*. Longman, London, pp. 56–59.

Thompson, K.F. (1976) Cabbages, kales etc. In: Simmonds, N.W. (ed.) *Evolution of Crop Plants*. Longman, London, pp. 49–52.

Wien, H.C. and Wurr, D.C.E. (1997) Cauliflowers, broccoli, cabbage and Brussels sprouts. In: Wien H.C. (ed.) *The Physiology of Vegetable Crops*. CAB International, Wallingford, Oxon, UK, pp. 511–552.

Chinese and other Asiatic cabbages

Herklots, G.A.C. (1972) Asiatic cabbages. In: Herklots, G.A.C. (ed.) *Vegetables in Southeast Asia*. George Allen and Unwin, London, pp. 190–224.

Opeña, R.T., Kuo, C.G. and Yoon, J.Y. (1988) *Breeding and Seed Production of Chinese Cabbage in the Tropics and Subtropics*. Technical Bulletin No. 17. Asian Vegetable Research and Development Center, Taiwan.

Radish

Shinohara, S. (1984) *Vegetable Seed Production Technology of Japan, Elucidated with Respective Variety Development Histories, Particulars*, Volume 1. Shinohara's Authorized Agricultural Consulting Engineer Office, Tokyo, Japan, pp. 195–237.

9 *Cucurbitaceae* – Cucurbits, the Vine Crops

The important vegetable crops in this family which are cultivated in the tropics include:

- *Citrullus lanatus* (Thunb.) Mansf. syn. *Citrullus vulgaris* Schrad. – watermelon, egusi;
- *Cucumis melo* L. – canteloupe, melon, sweet melon, honeydew;
- *Cucumis sativus* L. – cucumber;
- *Cucumis anguria* L. – West Indian gherkin, bur gherkin;
- *Cucurbita maxima* Duch. ex Lam. – pumpkin, winter squash, Chinese pumpkin, crookneck squash;
- *Cucurbita moschata* (Duch. ex Lam.) Duch. ex Poir – pumpkin, winter squash;
- *Cucurbita mixta* Pang. – pumpkin, winter squash;
- *Cucurbita pepo* L. – marrow, vegetable marrow, courgette, summer squash;
- *Cucurbita ficifolia* Bouche – Asian pumpkin, Malabar gourd, fig-leaf gourd, silacayote, vitoria;
- *Benincasa hispida* (Thunb.) Cogn. – ash pumpkin, Chinese watermelon, hairy wax gourd, tallow gourd, wax or white gourd;
- *Lagenaria siceraria* (Molina) Standl. syn. *L. vulgaris* Ser.: *L. leucantha* (Duch.) Rusby – bottle gourd, calabash gourd, white-flowered gourd, zucca melon;
- *Luffa acutangula* (L.) Roxb. – angled gourd, silky gourd, angled loofah, ridged gourd, vegetable gourd;
- *Luffa cylindrica* (L.) M.J. Roem syn. *L. aegyptiaca* Mill. – dishcloth gourd, smooth loofah, sponge gourd, vegetable sponge;
- *Momordica charantia* L. – balsam pear, bitter cucumber, bitter gourd, bitter melon;
- *Sechium edule* (Jacq.) Swartz – choyote, chayote, christophine; and
- *Trichosanthes cucumerina* L. syn. *T. anguina* (L) Haines – snake gourd.

The current thinking on the origins, nomenclature, descriptions and the distinguishing characters of the cucurbitaceous species has been outlined by Robinson and Decker-Walters (1997). The cultivated genera are frequently referred to as *cucurbits* and are widely distributed in the tropics. They are often referred to as 'vine' crops. All the genera are susceptible to frost. Many of the cucurbits are important crops especially in the drier regions of the tropics because they crop well under arid conditions with irrigation and also tolerate high temperatures and low relative humidity.

Some of the crops, for example watermelon and melon, are not regarded as vegetables but as 'fruit' by some authorities because they are eaten as a dessert. However, because of the production methods of these crops they are generally considered as vegetables from their agronomic aspect.

The majority of the *Cucurbitaceae* species have unisexual flowers borne on monoecious plants, an exception is *Telfairia occidentalis* Hook. f. (commonly known as fluted pumpkin or telfairia nut) which has unisexual flowers on separate plants.

Citrullus lanatus (Watermelon)

Origin and distribution

This crop is widely grown throughout the tropics, subtropics and arid regions of the world. It originates from Africa but there are references to the very early cultivation of the species elsewhere in the world (Hedrick, 1972). In some parts of Africa local cultivars with relatively bitter fruits are cultivated for their seeds which are roasted and eaten; these are usually referred to as 'egusi'.

Climatic requirements

The watermelon is a vigorous annual which covers a large area of ground with its sprawling stems; it can survive relatively dry conditions because of its deep root system. For this reason it has become established as an important crop in many tropical areas, especially where arid conditions prevail.

This species requires temperatures above 20°C for satisfactory growth and fruit development. Generally it thrives in a dry climate; wet or rainy conditions provide a suitable environment for foliar pathogens and may also result in a low sugar content of mature fruit.

Types and cultivars

Watermelons have been selected and improved by plant breeders, especially in the USA and parts of Asia, where a very wide range of cultivars is maintained. Some of the cultivars originally bred in the USA, for example 'Charleston Gray', are widely cultivated in tropical areas of the world. The *Citrullus* species are distinguished from the *Cucumis* species (cucumbers) and *Cucurbita* species (squashes) by their deeply pinnately lobed leaves. The plants have long, trailing vines up to approximately 4 m. Shinohara (1984) has described the history of cultivar development in watermelons.

Site and soil

Watermelons are very susceptible to *Fusarium* and should not be grown on the same site more frequently than every 4 years; this species should not follow another cucurbit crop but ideally follow cowpeas or maize in planned rotations. The optimum soil type is a well-drained sandy loam with high organic matter content.

The tolerable pH range is 5.0–6.8, although ideally 6.0–6.8. The addition of bulky organic manures during site preparation is very beneficial. As an alternative, a base fertilizer dressing of N:P:K 1:1:1 can be applied at a rate of 20 g/m^2 during the final site preparations. Higher nitrogen levels than this should be avoided otherwise the plants make excessive vegetative growth at the expense of flowers and fruit; the nitrogen values of organic materials applied during preparations should also be taken into account to avoid excess available nitrogen.

Crop establishment

Seeds are usually sown direct into the field in preference to raising plants in nurseries.

The plants are either grown on 'mounds', flat ridges or on the flat. The system adopted depends on the irrigation system to be used. Mounds or ridges are used in conjunction with a furrow or similar irrigation system. The seeds are spot sown, usually two to three seeds per station 90–120 cm apart in the row, with 120–180 cm between the rows. Seeds are slow to germinate when soil temperatures are low. If sown under dry soil conditions rats may take the seeds before germination. The plants are very frost sensitive and growers generally have a local 'rule of thumb' for a 'safe' sowing date after the likelihood of frost has passed. The seedlings are thinned out to their final stand when the first true leaves are showing.

The frequency of irrigation will depend on the soil type and climate, but because watermelon plants develop a deep and extensive root system water applications can be kept to a minimum; in dry regions sufficient irrigation should be applied prior to sowing to restore the soil to field capacity.

The crop should be weeded during the same operation as thinning the seedlings. The young plants tend to produce shallow roots in their early stages therefore care must be taken not to damage them. The crop is irrigated, usually on a weekly basis, although this depends on local conditions and availability of irrigation.

Table 9.1. Pests and pathogens of watermelon.

Pest or pathogen	Common names
Colletotrichum sp.	
Didymella bryoniae	Gummy stem blight, black rot
Fusarium oxysporum f.sp. *niveum*	Fusarium wilt
Glomerella lagenaria	Anthracnose
Squash mosaic virus	

Flowering and pollination

Watermelon is a day-neutral plant and there are therefore no problems in flower initiation. However, plant and fruit development are poor when ambient temperatures are less than 25°C during anthesis. Watermelon flowers are insect pollinated; mainly by honeybees therefore care must be taken not to adversely affect honeybees or other pollinating insects with pesticides. In exposed situations shelter from prevailing winds will improve the microenvironment for pollinating insects. The crop should be irrigated during anthesis.

Small developing fruit are usually thinned to leave two per vine but the vine is not pruned or reduced in length.

Harvesting

Time from sowing to harvest depends on the cultivar and its mature fruit size. An approximate rule of thumb is 3 months from seedling emergence or 5 weeks from anthesis. A useful indication of an individual fruit's maturity is the senescing stem tendrils at the base of the fruit.

Pests and pathogens

The pests and pathogens of watermelon are listed in Table 9.1.

Cucumis melo (Canteloupe, Melon, Sweet Melon, Honeydew)

Origins and distribution

The melons originated in tropical and subtropical Africa. There are also secondary centres of diversity in parts of Asia which include China, the Indian subcontinent and Iran.

Environmental requirements

Fairly dry conditions with a low relative humidity with temperatures between 24 and 28°C. Shelter from a strong prevailing wind will improve the micro-environment in exposed situations especially ensuring improved conditions for pollinating insects.

Soils

Soils which have a high organic content are satisfactory especially in areas likely to have low rainfall or insufficient available irrigation water. However, the crop succeeds very well on sandy soils which are free draining, on sites where there is a reliable supply of irrigation water during the cropping season.

The required soil pH range is 6.0–6.8. The traditional applications of bulky organic manures for this crop can be replaced by using rotations of forage crops, such as lucerne (alfalfa) (*Medicago sativa* L. and *Medicago* x *varia*) which improve soil structure.

Fertilizers applied during the final stages of seedbed preparation should have an N:P:K ratio of 1:1.5:2, applied at a rate of 50 g/m^2 although appropriate reductions of the phosphorus and potassium levels should be made to allow for residues which are already present in the soil or derived from any organic mate-rials incorporated during site preparation. Where leaching is likely to be signifi-cant some of the nitrogen can be applied as a top dressing at the start of anthesis, although nitrogenous fertilizers should not be applied once fruits begin to set as they will delay their ripening and maturity.

Crop establishment

The melon crop is grown on flat ground, ridges or raised beds in rows 1.5–2 m apart with 1.2–1.5 m between stations according to local custom and cultivar. It is normal practice to sow two to three seeds per station at a depth of 1–2 cm on the prepared site. After emergence the seedlings are thinned to one strong plant per station, weeding should also be done while hand thinning.

Some growers produce transplants in containers under protection and plant them out when they are large enough, at the same spacing as indicated above. The adoption of this method may depend on either cropping schedules or length of growing season.

The plant's leaders and developing laterals are 'stopped' when about four fruits have set per plant.

Top dressings of N:P:K in the same ratio as given above for site preparation should be applied when the young plants are established and again at the start of anthesis; if the pH was relatively low at the time of preparation a dressing of a liming material should be applied at the rate of 40 g/m^2 before irrigating, while the plants are still in the vegetative stage.

Fig. 9.1. Melons which are almost mature, indicating the position of an abscission layer when ripe, at the 'full slip' stage.

Harvesting

The first fruit are normally ready for harvest 3–4 months from seedling emergence or planting out.

The fruit of the cantaloupe and muskmelons tend to separate from the stem at the base of the fruit as the fruit becomes fully mature. This stage of separation by formation of an abscission layer between the stem and a mature fruit is usually referred to by melon growers as 'full slip'; this is indicated in Fig. 9.1. Other melon types are harvested after their characteristic external maturity colour is reached by cutting the stem from the mother plant just beyond 2 cm from the fruit.

Pests and pathogens

The pests and pathogens of *Cucumis* species are listed in Table 9.2.

Table 9.2. The main pests and pathogens of *Cucumis* species.

Pest or pathogen	Common names
Tetranychus cinnabarinus	Carmine spider mites
Thrips palmi	Melon thrips
Frankliniella occidentalis	Western flower thrips
Bemisia tabaci	Whitefly
Meloidogyne incognita	Root knot eelworm
Pythium spp. and *Rhizoctonia solani*	Damping off
Pseudoperonospora cubensis	Downy mildew
Erysiphe polyphaga	Powdery mildew
Alternaria cucumerina	Alternaria leaf spot
Colletotrichum lagenarium	Anthracnose
Didymella bryoniae	Leaf spot, black rot, gummy stem blight
Fusarium oxysporum	Fusarium wilt
Pseudomonas syringae pv. *lachrymans*	Angular leaf spot
Erwinia sp.	Bacterial soft rot
Cucumber green mottle mosaic virus (CGMMV)	
Cucumber mosaic virus (CMV)	
Zucchini yellow mosaic virus (ZYMV)	

Cucumis sativus (Cucumber)

Origins and distribution

It is generally considered that the cucumber originated in northern India, in the foothills of the Himalayas from where it was taken to secondary centres in China and the Near East then later further distributed in the world.

Cucumbers are widely cultivated in the tropics although development of improved cultivars has not received a very high priority due to the fruit's relatively low human nutrition value. The fruit is generally eaten raw; the young leaves are cooked as 'spinach' or consumed raw in parts of Asia.

Types

The majority of cucumber crops grown in the tropics are from trailing (vine) cultivars although bush cultivars are also available. Research and development originally aimed at the protected cropping industry in temperate areas, such as northern Europe, has resulted in cultivars which combine parthenocarpy and F_1 hybrids; this combination can provide growers with production of parthenocarpic fruit and resistance to serious soilborne pests and pathogens, such as root knot eelworm, Verticillium and Fusarium wilt diseases.

Site and soil

Cucumbers thrive best in a sheltered environment in preference to sites with significant prevailing winds. The majority of tropical crops are grown on the soil surface. However, some growers prefer to produce the crop vertically on a trellis which is approximately 3–4 m high consisting of netting (either plastic or wire) supported by vertical posts approximately 4–5 m apart. The total weight of the crop's vine, foliage and fruit (also potential water on leaf surfaces) must be taken into account when finalizing the dimensions of the support system in addition to selection of suitable locally available materials. An example of this type of overhead trellis erected for cucumber and other cucurbits with a similar morphology is illustrated in Fig. 9.2. Some subsistence farmers sow cucumbers adjacent to the supports of an indeterminate tomato crop which is coming to an end; this takes advantage of an existing support system.

The optimum soil pH is 5.5–7.0. Soils with satisfactory water retention are suitable provided they have efficient drainage. The crop responds to soils with high organic matter content, therefore if decomposed organic manure is available it should be incorporated during the early stages of site preparation.

A suitable base application of N:P:K nutrient ratio of 1:2:2 should be applied during the final stages of seedbed preparation at a rate of approximately 30 g/m^2 but the nutrient value of any bulky organic manures already applied should be taken into account. The amount of nitrogen is increased on soils with a high phosphorus and potassium status. A higher proportion of nitrogen is also necessary where frequent irrigation is required in order to allow for leaching; in this situation the N:P:K ratio should be nearer to 2:1:1, with approximately half the nitrogen applied as a top dressing approximately 1 month after seedling emergence. Care should be taken to avoid foliar scorch from this application.

Fig. 9.2. Overhead trellis structure for climbing cucurbits with pendulous fruit.

Crop establishment

When not produced on upright supports, the crop is grown on prepared ridges, mounds or on the flat. The cucumber crop is produced on flat ridges or raised beds if irrigation is likely to be required. Seeds are sown approximately 2 cm deep at stations approximately 12 cm apart in rows up to 2 m apart. After seedling emergence the young plants are thinned to approximately 30 cm apart in the rows.

Harvesting

The field types cultivated in the tropics reach maturity 15–20 days from pollination. The individual plants are very capable of continuous cropping over a period of approximately 12 weeks, although this will depend on the local climatic conditions, cultivar and absence of pests and pathogens which affect productivity and longevity of the mother plant. In areas with suitable climatic conditions further sowings can be made at 2 monthly intervals to maintain continuity of supply.

Pests and pathogens

The pests and pathogens of *Cucumis* species are listed in Table 9.2.

Cucumis anguria (West Indian Gherkin, Bur Gherkin)

This species is locally important in South America, especially Brazil, and is also important in the West Indies.

It is an annual with a trailing habit, the immature fruit which have a rough and spined rind are harvested when they are approximately 6 cm long and 4 cm wide; the fruit are reminiscent of gooseberries in external appearance (i.e. fruits of *Ribes grossularia* a north European cultivated bush fruit). The harvested *C. anguria* fruit are cooked, used in curries, eaten raw or pickled. This species is not the small-fruited form of *C. sativus* which is frequently referred to as gherkin.

Cultivation

The crop is produced from seed as described for *C. sativus* and the crop is produced on the flat in the same way.

Pests and pathogens

The main pests and pathogens of *Cucumis* species are listed in Table 9.2.

Marrows, Pumpkins and Squashes

This group of cucurbits contains the wide range of types within *Cucurbita pepo.* Robinson and Decker-Walters (1997) have discussed and described the groups of cultivars considered to be within this species. The *Cucurbita* species discussed here in this section are:

- *C. maxima* – pumpkin, winter squash, Chinese pumpkin, crookneck squash;
- *C. moschata* – pumpkin, winter squash;
- *C. mixta* – pumpkin, winter squash; and
- *C. pepo* – marrow, vegetable marrow, courgette, summer squash.

Origins and distribution

The *Cucurbita* species originated in the southern area of North America, Mexico and down to northern Argentina. There are secondary areas of diversity for *C. maxima* in India; there is also evidence of very early introductions into South-east Asia (Grubben, 1977).

As can be seen from the common names within this group of cucurbits there is great difficulty in defining the different cultivars; taxonomists continue to clarify the situation.

The species can be further defined according to their season of production and/or the time of year each is used for culinary purposes; a rule of thumb classification is:

- Pumpkin – the fruit of any of the four species listed above which is used as a vegetable or in some situations used as a stock feed.
- Summer squash – edible fruit of any of the above four species but those which are used for culinary purposes while still immature, especially *C. pepo.* An example of a summer squash, cv. Patty Pan, is illustrated in Fig. 9.3(a). Examples of winter squash morphological types are illustrated in Fig. 9.3(b).
- Winter squash – edible fruit or any of the above four species which have a mild flavour and fine-textured flesh. When left on the mother plant the rind ripens to produce a fruit which can be stored and used for culinary purposes from storage or used as stock feed.

Soil pH and nutrition

This group of *Cucurbita* species is moderately tolerant of acid conditions and can be produced successfully on soils with a pH from 5.5 to 6.8.

These vegetables respond to applications of bulky organic manures, if available, during site preparation, although reasonable crops can be produced by applications of inorganic fertilizers. The optimum N:P:K ratio is 1:2:2 with appropriate deductions made for nutrients applied as bulky organic materials at a rate of 60 g/m^2 followed by a top dressing of N:P:K 1:1:1 at a rate of 30 g/m^2 when

Fig. 9.3. (a) Summer squash cv. Patty Pan (photograph courtesy of The Real IPM Company (K) Ltd; www.realipm.com). (b) Morphological types of winter squash.

the first fruits start to set. On soils with a high P and K status top dressings of only nitrogen are used. The nitrogen top dressing is particularly important when leaching has occurred.

Crop establishment

There are three methods of field production: (i) on the flat; (ii) on flat ridges; or (iii) on mounds (the latter are sometimes referred to as 'hills'). The system adopted depends on the irrigation system and efficiency of soil drainage on the site as this crop is not successful on waterlogged soils. The preparation of mound systems is very labour intensive as it is difficult to mechanize the operation but can be suitable on a very small scale.

The ultimate planting density depends on the type of cucurbit and whether it has a trailing ('vine') or bush growth habit. Row spacings of 90 cm to 3.5 m are used according to the vigour of the cultivar to be grown. The shorter width between rows is used for the less vigorous and bush cultivars, whereas the wider row widths are used for the more vigorous trailing types. The distance between plants within the rows is usually the same as the distance adopted between rows. It is normal practice to sow about three seeds per station and single out the seedlings (i.e. reduce seedlings to one per station) after emergence unless the seeds are sown with a precision drill but this would require seeds of a high and reliable potential germination rate.

Pollination is essential for satisfactory fruit set and subsequent development, and pollination is generally by bees although some other insect species are known to pollinate *Cucurbita* flowers. There is a high possibility of cross-pollination with some other *Cucubita* species therefore the grower would be well advised to use a reliable seed source for further crops in preference to saving their own seed. Saving seed from hybrid cultivars is also undesirable as the progeny will be segregating and will very likely produce crops which are not true to type or not be of significant culinary value.

Harvesting

All squashes, pumpkins and marrows take approximately 16 weeks from anthesis to seed maturity. By this stage the rind has hardened and usually changed colour. The green types change to a yellow-orange colour and the yellow-golden types change to a straw colour. Fruit of the winter squash, long-storing types, including the Hubbard's types, should be exposed to the sun before storing in order to harden the rinds. A short length of stem should be left on the fruit when cutting from the mother plant.

Pathogens

The main pathogens of *Cucurbita* species (pumpkin, vegetable marrow and squash) with common names of the diseases they cause are listed in Table 9.3.

Table 9.3.   The main pathogens of *Cucurbita* species.

Pest or pathogen	Common names
Cladosporium cucumerinum	Scab
Didymella bryoniae	Leaf spot
Fusarium solani f. sp. *cucurbitae*	Fusarium foot rot
Xanthomonas campestris pv. *cucurbitae*	Bacterial leaf spot
Cucumber mosaic virus (CMV)	
Mosaic musk melon virus	
Squash mosaic virus	

Cucurbita ficifolia (Asian Pumpkin, Malabar Gourd, Fig-leaf Gourd, Silacayote, Vitoria)

Origin and distribution

This *Cucurbita* species originates in South America where it has been a well-established food crop in the higher altitudes of Mexico and more southerly areas of South America between 1000 and 3000 m. There is archeological evidence from Huaca Prieta, Peru, suggesting that cultivation of this species dates back to 2000 BC (Herklots, 1972). It is now widely cultivated in South America for sweet and conserve preparations, also in the Philippines and some other parts of Asia. The seeds are high in fat and protein levels and are widely used as a food. Mature fruits can reach weights of up to 5 kg and along with the vines are often used for stock feed. The flowers, immature fruit and young shoots are cooked as vegetables.

Cucurbita ficifolia is a perennial with sprawling vines whose tendrils enable it to scramble over scrub. According to Herklots (1972) this species requires short days for flowering. Although it is a perennial species, it is only propagated from seed. Interestingly, it has been used as a root stock for other cucurbits to provide resistance to some soilborne pathogens, especially in the protected cropping industries of temperate climates such as northern Europe.

Cultivars and types

There are no reported cultivars although the mature seed's testa can range in colour from pale brown to black in seeds harvested from separate plants.

Cultivation

Cucurbita ficifolia is cultivated as described above for *Cucurbita* species.

Pests and pathogens

The main pathogens of *C. ficifolia* are those of other *Cucurbita* species listed in Table 9.3.

Benincasa hispida (Ash Pumpkin, Chinese Watermelon, Hairy Wax Gourd, Tallow Gourd, Wax or White Gourd)

Origin and types

This is the only cultivated species of the genus; it is thought that it originates from Java (Indonesia) and possibly also from Japan. It is an important crop in parts of Asia, especially the tropical areas. The mature fruit can be stored for many months (sometimes up to a year) without refrigeration which is one of its attributes and main culinary importance; the young leaves and inflorescences may be cooked as a green vegetable. Most seed stocks are based on local selections and there are also some named cultivars. Up to three crops can be produced in some areas each year.

Cultivation

The crop requires a well-drained site with available water. The usual preparation is to partly fill an open trench with decomposed manure or compost and backfill it with the top soil which is then firmed by treading; in some situations the organic material is spread in a band approximately 1 m wide and covered in top soil to a depth of up to 40 cm. This latter method is preferable where drainage is impeded and the crop is to be produced using an existing fence or other structure for support. The crop can be grown on the flat and because of its vigour it scrambles over the ground and any shrubby plants it encounters; according to Herklots (1972) shoots have been reported as growing an average of 1 inch (approximately 2.5 cm) every 3 h. The crop is more commonly grown on supports but these must be substantial, the sides of sheds or other structures with trellises can be used for this purpose. If fruit is not setting the female flowers should be hand pollinated. The fruit which are to be left on the plant to reach their maximum potential size should be individually supported by nets to ensure that they do not become mechanically damaged or damage the vine by tearing off.

Harvesting

Some partly developed fruit may be harvested for culinary purposes. The mature fruit is usually harvested approximately 5 months from plant establishment although shoots, inflorescences and immature fruit can be harvested earlier. The mature fruit is cylindrical and 1 m, or more, in length and at maturity can weigh from 10–45 kg. The fruits are covered in a hairy white wax when mature. The mature fruits are stored on shelves (ideally slatted) in a cool shed or building. In some areas they are stored in cellars.

Pests

The pests of *B. hispida* are:

- *Aphis gossypii* – cotton aphid; and
- *Dacus* spp. – fruit flies.

Lagenaria siceraria syn. *L. vulgaris*: *L. leucantha* (Bottle Gourd, Calabash Gourd, White-flowered Gourd, Zucca Melon)

It should be noted that this species should not be confused with the true calabash which is the fruit of *Cresentia cujete* L., a small tree in the family *Bignoniaceae*.

Origins and types

According to Purseglove (1974) 'this species is probably a cultigen which was common to both the Old and New Worlds since remote times' and that 'the bottle gourd originated in Africa or India where it occurs subspontaneously, as it also does in India'.

The species is now widely cultivated in the tropics as a food crop and for fashioning containers, utensils and ornaments produced from its hard and ripened fruit skin. Some selections are relatively bitter, but less bitter types and cultivars are available for culinary purposes.

Crop production

Lagenaria species are very suitable crops for high temperature and low relative humidity environments. A 2–3 year crop rotation should be exercised on sites which have a history of Fusarium wilt. A reliable seed supply should be used as far as possible because this wilt is seed transmitted in addition to being soilborne. Preparation and sowing are similar to those described for the squashes except that *Lagenaria* has to have vertical stakes for plant support. The final spacing of single plants is approximately 3 m each way. This species has tendrils which easily attach themselves to the supports and should normally not require tying in. Overhead trellis (horizontal) which bridges the vertical supports can ensure that the developing fruit hangs down, producing a high quality fruit as illustrated in Fig. 9.2. However, when produced in very dry areas or dry seasons the supports can be dispensed with to reduce labour and materials by producing the crop on the flat.

Although insect pollinated, it is customary to hand pollinate the female flowers if satisfactory setting is not observed. The growing points should be 'stopped' (nipped off) when the vines are 3–4 m high. Fruit which are likely to touch the ground if left to mature should be removed early to avoid soil contact which is likely to result in fruit rot; when grown on the flat a tile or similar material should be placed under each developing fruit.

Harvesting

The young shoots, young foliage and immature fruits are cooked and eaten from those cultivars which have been developed or selected with reduced bitterness. The seeds derived from mature fruits are used in soups. Those types destined for containers or ornaments are taken from the mother plant when fully mature and usually further dried off, ideally in a dry atmosphere. The storage life of fruits harvested for food before rind ripening is relatively short and, subject to being free of mechanical damage and disease, fruit can be stored in cool conditions for approximately 2 weeks.

Pests and pathogens

Lagenaria species are prone to:

- anthracnose; and
- Fusarium wilt.

Luffa Species (Gourds and Loofahs)

There are two important vegetable species of *Luffa*:

- *L. acutangula* – angled gourd, silky gourd, angled loofah, ridged gourd, vegetable gourd; and
- *L. cylindrica* syn. *L. aegyptiaca* – dishcloth gourd, smooth loofah, sponge gourd, vegetable sponge.

Origin and types

There are groups of cultivars suitable for vegetable production and also groups of those cultivars best suited for loofah ('sponge') production. Although *L. cylindrica* is the species cultivated for production of the 'sponge' of fibrous material obtained from the mature and treated ripe fruit, young shoots and leaves of both species are harvested for use as a cooked vegetable. There is a range of bitterness in the *L. cylindrica* cultivars, thus only the more palatable ones are used as vegetables.

The vegetable loofahs are widely grown in the tropics but are most important as a vegetable crop in South-east Asia including India. Figure 9.4 shows a crop of angled gourds in Malaysia.

Cultivation

Site preparation for both species is as described for the squashes, that is on ridges or mounds in rows 1.5 m apart. The *Luffa* vines are sufficiently vigorous to compete with weeds, although some surface cultivation may be necessary until the plants have become established.

The vegetable *Luffa* crop is usually provided with supports, as illustrated in Fig. 9.2 although in the dry tropics it can be produced as a ground-cover crop. Irrigation is usually required during anthesis, especially during periods of low rainfall.

Harvesting

The vegetable *Luffa* cultivars are generally harvested for their young shoots and leaves as required once the plants are established. Also the immature fruit

Fig. 9.4. A crop of angled gourds growing in Malaysia, note the strong supports and black polythene mulch.

are harvested as required. Fruit should be harvested before the internal fruit fibres develop. As a rule of thumb this is not more than half the fruit's potential mature size.

Pests and pathogens

Pests and pathogens of *Luffa* are:

- *Dacus* spp. – fruit flies;
- *Pseudoperonospora cubensis* – downy mildew; and
- *Erysiphe chicoracearum* – powdery mildew.

Momordica charantia (Balsam Pear, Bitter Cucumber, Bitter Gourd, Bitter Melon)

Origins and types

The genus *Momordica* has several species which are either cultivated, escapes or indigenous from which the fruit and/or shoots are gathered as a food source. *Momordica charantia* is by far the most important species in cultivation. It is thought to have originated in the Indian subcontinent with secondary diversity in South-east Asia, China and Africa. It is believed that the species reached South America via the slave trade. The crop is now widely produced in China, South-east Asia and the Indian subcontinent, Africa and South America.

There are many local selections in addition to cultivars maintained by seed organizations. The selections and cultivars vary in the level of bitterness of their fruit and there are also long- and short-fruited cultivars.

The young and bitter fruits are used as vegetables. The fruits are pickled or used in curries and in some areas the young vine shoots and leaves are used as a cooked green vegetable.

The crop is very useful, especially for subsistence farmers and their dependents, for its relatively high levels of iron and vitamin C.

Cultivation

The species is a slender annual climber, normally grown to a height of approximately 2–3 m on vertical supports; cut bamboos made into a trellis are widely used in Asia. The crop grows well in the hot humid environments, ideally on soils with good water retention which have had organic material incorporated during site preparation.

The crop is grown in rows with two to three seeds sown at stations approximately 60 cm apart. The seedlings are singled out to one per station following their emergence.

A top dressing of well-composted organic material can be applied following plant establishment otherwise a top dressing of a nitrogenous fertilizer can be applied at a rate of 34 g/m run of bed if growth is slow or weak.

The fruit fly, *Strumeta cucurbitae*, is a common pest in many areas. Some growers protect the pendulous developing fruit by enclosing it in a paper cylinder which is secured by tying string around the peduncle (fruit stalk).

Harvesting

Immature fruit may be harvested from approximately 7–9 weeks from seedling emergence depending on local conditions and the desired state of development for specific culinary preparations. The more mature fruits are generally harvested when their rinds start to turn a paler green to yellow. The mature fruit cannot be stored for more than a few days unless a low temperature store at 2°C is available.

Sechium edule (Choyote, Chayote, Christophine)

Origin and types

This is a perennial climber which is indigenous to Mexico and Central America but now widely cultivated throughout the tropics including parts of Africa, India and South-east Asia.

The fruits, and sometimes tuberous roots, are used as a cooked vegetable although culinary use of the root is more closely associated with Asia. This crop is common and relatively easy to grow on subsistence farms and further development and selection could increase its value for cultivation. Many of the selections used are known by alternative morphological characters such as 'white' or 'green', 'pointed' or 'blunt', 'spiny' or 'smooth', ' round' or 'long' (Grubben, 1977).

Cultivation

Planting stock and selections are usually made by vegetative cuttings from plants with optimum fruit and suitable morphological characters. Each fruit bears a single seed; the seed is considered to be recalcitrant and is usually kept in the intact fruit until the following planting season if propagation from seed is required.

The crop is normally grown adjacent to a trellis or another substantial support system which is approximately 2 m high. The entire fruit which contains a single seed is planted horizontally at a depth approximately the width of the fruit, covered with soil and firmed in. The young plants or seeds are planted along the supports at approximately 120 cm apart. If it is to be grown as a ground-cover crop the plant density should be halved per unit area.

Harvesting

Mature fruit can be harvested approximately 12 weeks from sowing or planting; fruiting continues over several months. The harvested fruit may be stored for several weeks.

Trichosanthes cucumerina syn. *T. anguina* (Snake Gourd)

Origin and types

This species is thought to have originated in India but is now cultivated in the humid tropical areas of Asia, Australasia, South America, the West Indies and locally in parts of Africa. It is a climbing annual cultivated for its immature fruit, young shoots and leaves which are cooked; the mature fruit is very bitter, very fibrous and not normally used for food.

There has been very little plant breeding development of this species; growers usually have their own seed from selected plants. However, breeding and development work is reported to be in progress at some Indian plant breeding institutes.

Cultivation

The crop usually succeeds in tropical areas with both high humidity and rainfall. It requires a soil with good water-holding capacity but also satisfactory drainage. In some areas the seeds are sown on each side of a raised bed with irrigation channels running along each side of the beds.

Soil preparation includes incorporating compost prior to sowing. If suitable compost is not available a general compound fertilizer of N:P:K 7:7:7 can be incorporated during soil preparation at a rate of 60 g/m^2. The crop requires vertical supports in the form of a trellis or bamboo canes, and overhead horizontal supports improve the facilities for the fruit to hang down. Some growers tie a weighted string around the tip of some fruits to keep them straight.

Several seeds are sown 1.5 cm deep, at stations approximately 60 cm apart in rows which are 1.5 m apart. The seedlings are singled out to one per station after they emerge.

Harvesting

The first fruit may be harvested approximately 2 months from sowing. In some areas the fruit may be stored for up to 2 weeks provided it is kept at a high humidity in temperatures of approximately 15–18°C.

Pests and pathogens

Pests and pathogens of *T. cucumerina* are:

- *Dacus* spp. – fruit flies;
- *P. cubensis* – downy mildew; and
- *Colletotrichum lagenarium* – anthracnose.

Further Reading

Herklots, G.A.C. (1972) *Vegetables in South-East Asia.* George Allen and Unwin London, pp. 273–348.

Robinson, R.W. and Decker-Walters, D.S. (1997) *Cucurbits.* CAB International, Wallingford, Oxon, UK.

Shinohara, S. (1984) *Vegetable Seed Production Technology of Japan, Elucidated with Respective Variety Development Histories, Particulars,* Volume 1. Shinohara's Authorized Agricultural Consulting Engineer Office, Tokyo, Japan, pp. 318–426.

10 *Solanaceae* – Tomatoes and Related Crops

The genera in this botanical family provide some of the most important vegetables in the tropics and include:

- *Lycopersicon lycopersicon* (L.) Karsten ex Farw. syn. *L. esculentum* Mill. – tomato;
- *Capsicum annuum* L. – sweet peppers, cayenne peppers, chillies;
- *Capsicum frutescens* L. – chilli pepper;
- *Solanum melongena* L.– aubergine, brinjal, eggplant; and
- *Solanum tuberosum* L. – potato, Irish potato, white potato.

Other economically important species in this botanical family include *Nicotiana tabacum* L. (tobacco) and some other poisonous alkaloid-bearing drug plants including *Atropa belladona* L. and *Datura stramonium* L. Although these last three species are not vegetables, it is important to appreciate that they are related crops which are alternative hosts and sources of the viruses which infect a wide range of genera in the *Solanaceae*; many of the viruses are transmitted by a range of vectors including insects, implements and workers.

Lycopersicon lycopersicon syn. *L. esculentum* (Tomato)

Origins and types

The primary centre of origin of *L. lycopersicon* is Mexico and the mountains of the American West. Several *Lycopersicon* species, including *L. cheesmanii* Riley, *L. pimpinellifolium* Mill., *L. chilense* Dun. and *L. hirsutum* Humb. are important species used in plant breeding programmes as sources of resistance to some economically important pests and diseases of the cultivated tomato. The development of F_1 hybrids by plant breeders has provided cultivars with resistance to specific eelworm species as well as serious soilborne pathogens.

Two morphological types of *L. lycopersicon* are cultivated: (i) determinate; and (ii) indeterminate. Cultivars of both types are available with resistance to specific pests and pathogens as a result of plant breeding and increased availability of improved seed.

The determinate cultivars are commonly referred to as *bush tomatoes*. The fruit and foliage of these cultivars can be more prone to some pests and pathogens especially in areas of high relative humidity or heavy rainfall.

The indeterminate cultivars are referred to as cordon or single stem cultivars, requiring removal of side shoots and being grown on a system of supports, stakes or vertical trellis. The supports are usually procured from local materials. These cultivars generally have a higher labour requirement but can generally be expected to produce better quality fruit and are also capable of producing fruit over a longer period; however, they can be more susceptible to sunscald.

Environmental requirements

Tomatoes produce a satisfactory crop and yield when grown within the temperature range of 20–26°C. The plants will survive at temperatures higher than this but they will not normally produce viable pollen and therefore fruit set will either be poor or absent.

Soil and site

There should be at least a 1 year break between successive tomato crops, ideally 2 or more years in order to minimize any build up of soilborne eelworms and soilborne pathogens. This break requirement also includes *S. melongena* (eggplant) and *Capsicum* spp. (peppers). Suitable crops during the break are groundnut, rice or a fodder depending on the crops normally produced and required by the grower.

The optimum pH is 6.5–7.0. Although the crop may often succeed at slightly lower pH levels there is a risk of calcium deficiency below pH 6.5. A symptom of calcium deficiency is 'blossom end rot' also called 'black spot'; fruit with this symptom are shown in Fig. 10.1. If calcium deficiency is the cause the symptom will generally occur on all successive ripening fruit; the deficiency can usually be corrected by a surface application of a liming material at a rate of $40\,g/m^2$ and watering it in. This same symptom on the fruit may also be the result of water shortage in the plant at the early development stage of the affected fruit following pollination. In this case it is usually only a small number of consecutive developing fruit on one truss and this can usually be prevented by improving the irrigation programme by ensuring that the plants are not allowed to become dry throughout anthesis.

Soil preparation

Well-decomposed organic manure or compost can be incorporated during site preparation, but animal manure which has not completely decomposed can

Fig. 10.1.   Field-grown tomato fruit showing symptoms of blossom end rot.

cause root scorch from free ammonia and nitrites. This is especially important in seedbed preparation.

Any inorganic fertilizers applied during preparation should take account of existing soil nutrient level values included in the organic materials. In principal the optimum N:P:K levels are 1:2:2 and these should be incorporated at the rate of $60\,g/m^2$ of bed. If the available nitrogen is too high, excess foliage can be produced at the expense of earliness of ripe fruit. If this is observed it should be counteracted by light top dressings of a high potassium fertilizer such as potassium sulfate at the rate of $30\,g/m^2$. Conversely, if growth rate is slow and no other pathological reason or explanation can be determined, then a light top dressing of a soluble nitrogen fertilizer such as ammonium sulfate can be applied at the rate of $20\,g/m^2$.

Plant production

Tomato seedlings are most susceptible to a range of pre- and post-emergence damping-off diseases until they have their first true leaves (approximately 2 weeks from seedling emergence). It is important that seed has been obtained from a reliable source and is as far as possible free from seedborne pathogens. Transplants are usually raised in outside seedbeds with good drainage and are planted in their final quarters when large enough to handle. In some situations, where there is a short growing season for tomatoes, seedlings are raised in containers under plastic or other protection and planted out in their final positions either when large enough or when the weather conditions are satisfactory. Some protection by way of a transparent covering should be provided for seedbeds when heavy rainfall is expected. Protection from strong, dry or dust-laden winds is often an advantage in exposed areas. Protection for raising young transplants as illustrated in Fig. 5.6 can also be useful over tomato seedling beds.

Mulching

A fine mulch, such as chopped rice straw applied to the seedbeds after sowing will help to suppress weeds and also protect the soil surface from damage by heavy rains.

Mulches can also be applied to the transplanted crop immediately following planting or later, as the plants become established, this would depend on the friability of the mulching material. Black polythene sheeting is not always readily available to small farmers either because of its cost or absence in the local markets; however it can be very useful for determinate and indeterminate cultivars grown on the flat and can be positioned and weighed down immediately prior to planting out. When black polythene is used it is put in position and secured around the edges before planting takes place. Mulching also protects fruit from water and soil splashing and is especially useful in this respect for the bush cultivars, but for both bush and cordon tomatoes it can significantly reduce water lost by evaporation from the soil surface (see Fig. 10.2).

Planting determinate (bush) cultivars

The seedlings are usually planted out when they have developed their first two true leaves, and are approximately 8–10 cm tall by this stage. The furrows are taken out in preparation prior to planting when furrow irrigation is going to be used. The rows are 1–2 m apart, the wider spacing used for the more vigorous bush cultivars. The planting distance within the rows is 50–70 cm; the wider spacing is again used for the more vigorous bush cultivars.

Fig. 10.2. Cordon tomatoes growing with a polythene mulch, note also the symptoms of bacterial wilt on the lower foliage.

Planting indeterminate (cordon) cultivars
These are staked or provided with other forms of support and are planted in rows 70–90 cm apart and 45 cm between plants within the rows. Once the supports are in position and the plants have established, the removal of side shoots should start and continue on a regular basis, the main stems should be tied into the supports during the same operation.

Weeds should be controlled if mulches are not available, especially in the early stages of crop establishment; but care should be taken not to disturb or damage the tomato roots which can be near the soil surface.

Irrigation

The frequency of irrigation applications will depend on the prevailing rain pattern, but tomatoes require relatively regular watering either from rain, irrigation or a combination of the two. There should be a uniform level of available soil water especially from the time of anthesis. Irregular water availability can easily result in fruit disorders including blossom end rot ('black spot'), fruit cracking and fruit splitting (see Figs 6.3 and 10.1).

Harvesting

There is a general correlation between fruit size and earliness; as a rule of thumb the smaller fruited cultivars produce their first ripe fruit before the larger fruited cultivars. Generally the determinate (bush) cultivars start to produce ripe fruit 2 months from planting out while the indeterminate cultivars can take 9–10 weeks from planting out; however, these periods to first ripe fruit will very much depend on the cultivar and its interactions with the environmental conditions during plant growth and fruit development. Fresh fruit can be stored for several days in cool conditions but for long-term storage the fruit is usually 'sun-dried' (see Fig. 10.3) when the ripe fruit is sliced equatorially, placed on trays and sun-dried on the roofs of dwellings in the summer season.

Pests and pathogens

The main pests and pathogens of tomato are listed in Table 10.1.

Capsicum annuum (Sweet Pepper) and *Capsicum frutescens* (Chilli Pepper, Hot Pepper)

Origins and types

The peppers which are grown as vegetables are *Capsicum* species, not to be confused with *Piper nigrum* L., which is in the botanical family *Piperaceae*, also

Fig. 10.3.   Sun-drying ripe tomatoes on roof tops at Karimabad, Northern Areas of Pakistan.

Table 10.1.   Pests and pathogens of tomato.

Pest or pathogen	Common names
Liriomyza strigata	Serpentine leaf miner
Heliothis armigera	Corn ear-worm
Meloidogyne spp.	Root knot eelworm
Corynebacterium michiganense	Bacterial canker
Pseudomonas solanacearum	Bacterial wilt
Alternaria solani	Early blight, Alternaria blight
Cladosporium fulvum	Leaf mould
Corticium rolfsii	Seedling damping off
Fusarium oxysporum	Fusarium wilt
Verticillium albo-atrum	Verticillium wilt
Verticillium dahliae	Wilt
Pyrenochaeta lycopersici	Corky root, brown root rot
Phytophthora capsici	Seedling damping off
Phytophthora infestans	Late blight
Phytophthora nicotianae var. *parasitica*	Seedling damping off
Pythium aphanidermatum	Seedling damping off
Pythium myriotylum	Seedling damping off
Rhizoctonia solani	Soil rot
Sclerotium rolfsii	Quick wilt
Septoria lycopersici	Septoria leaf-spot
Stemphylium solani	Grey leaf-spot
Thanatephorus cucumeris	Seedling damping off
Tobacco mosaic virus (TMV)	

commonly called 'pepper', the fruits of which are used as a spice or condiment, including the production of 'black pepper' and 'white pepper'.

The two *Capsicum* species which are generally considered important as vegetables are *C. annuum* and *C. frutescens*. It is the level of capsicine in the fruit which is responsible for the 'hot' condiment effects of both species.

Capsicum annuum is an annual with flowers and fruits borne singly. This group forms a polymorphic cultigen and is capable of freely cross-pollinating within the group. Types of cultivar within this species are:

- sweet peppers (also known as bell peppers) which are red, green or yellow when ripe;
- cayenne peppers; and
- chillies.

Capsicum annuum originates from Meso-America and the Andes, and secondary centres have become established in parts of Europe, Africa and Asia. It is an annual. There are F_1 hybrids of the sweet and chilli types in addition to the open-pollinated cultivars.

Capsicum frutescens is a shrubby perennial with several flowers on each inflorescence, although it is frequently grown as an annual to avoid build up of viruses and severe foliage deterioration resulting from leaf and stem diseases, especially when produced under humid conditions. Cultivated types include the bird's eye chilli, cherry capsicum and the cluster pepper. Figure 10.4 illustrates a bird's eye chilli plant with ripening fruit.

Fig. 10.4. A bird's eye chilli plant illustrating ripening fruit. Photograph courtesy of The Real IPM Company (K) Ltd (www.realipm.com).

The *Capsicum* peppers originated in Meso-America and spread into the New World tropics before subsequent introduction into Asia and Africa. They are now widely grown throughout the tropics, subtropics and warmer temperate regions of the world; they are not frost tolerant.

The physiology of the cultivated *Capsicum* peppers has been reviewed by Wien (1997).

Environmental requirements

Both species are warm season crops which can be successfully cultivated at altitudes of up to 1700 m provided that irrigation is available. The ideal temperature range for both species is 21–27°C. Although low humidity reduces the incidence of foliar diseases, a high humidity in conjunction with the optimum temperature range provides the ideal conditions for flower and fruit development.

Site and soils

Rotation is an important consideration with these two crops; ideally at least 2 years should have lapsed between solanaceous crops on the same site, especially if root knot eelworms have been found previously on a related crop. Suitable break crops are rice, sorghum, millet or maize. However, peppers do not succeed in poor drainage; rotation with paddy rice means that the peppers are produced after the rice crop at the time of year when the fields are not flooded, but allowed to drain naturally. Poor drainage can result in leaf drop and potential crop reduction. The optimum soil pH is 6.0–6.5, if found to be lower than this a suitable liming material should be applied during site preparation at the rate of 30 g/m^2. Care should be taken in the application of base dressings as excessive nitrogen tends to delay fruit maturity. The optimum ratio of N:P:K applied is 8:6:8 incorporated as a base dressing at the rate of 30 g/m^2 during the final stages of site preparation followed by a similar quantity applied as a top dressing between the start of anthesis and fruit setting. Top dressings which only consist of a nitrogenous fertilizer should be avoided as they tend to promote vegetative growth at the expense of flower development and fruit set.

Plant production

Seedlings are usually produced in prepared beds, as described for tomatoes, or raised singly as transplants in containers.

Crop establishment

The young plants are transferred to their final positions when they are approximately 10–12 cm high. They are usually planted out in rows 60–80 cm apart and

35–45 cm apart within the rows, depending on the vigour of the cultivar; the *C. frutescens* cultivars are grown at the wider spacing. Where furrow irrigation is planned the ridges are prepared during site preparation. In areas where the climatic conditions are ideal, growers sow direct at the same inter- and intra-row spacing and in this method three or four seeds are sown at each station. The emerged seedlings are subsequently thinned to one strong seedling per station.

Cultural requirements

The uniform availability of soil water is important in order to ensure a continuous and sequential development of flowers and their fertilization. The crop should be kept free of weeds in the early stages of plant establishment.

Harvesting

With both species it is important to harvest each fruit with its stalk, this avoids rotting or premature deterioration at the fruits' calyx ends.

For sweet pepper, depending on cultivar, the first fruits can be harvested 7–11 weeks from transplanting, and can continue for a further 8 weeks.

For chilli pepper, depending on cultivar, the first chilli pepper fruits can be harvested approximately 6 weeks from transplanting. This species may be cultivated through a second year, although some deterioration in cropping ability may occur as a result of pests and pathogens.

The chilli peppers are usually sun dried, they subsequently store quite well provided that they are not kept in a high humidity atmosphere.

In some households a few plants are maintained as perennials for immediate culinary use; but it should be appreciated that these few perennial plants can be a pool for viruses and other pathogens.

Pests and pathogens

The main pests and pathogens of *Capsicum* spp. are listed in Table 10.2. Bacterial wilt symptoms are illustrated in Fig. 10.5.

Solanum melongena (Aubergine, Brinjal, Eggplant)

Origins and types

This species originated in India and was subsequently spread to Spain and Africa by Arab and Persian traders, respectively, and it also became an established crop in China. It is now very widely cultivated throughout the tropics.

There is a wide diversity of morphological types, especially regarding plant habit, fruit shape and size. Fruit colour between stable lines, or cultivars, ranges

Fig. 10.5. Chilli plant displaying symptoms of bacterial wilt. Photograph courtesy of The Real IPM Company (K) Ltd (www.realipm.com).

Table 10.2. The main pests and pathogens of *Capsicum* species.

Pest or pathogen	Common names
Ceratitis capitata	Fruit fly
Scirtothrips dorsalis	Thrips
Alternaria spp.	Fruit rot
Cercospora capsici	Frog-eye leaf spot, fruit stem-end rot, leaf spot
Colletotrichum capsici	Fruit rot
Colletotrichum piperatum	Ripe rot, anthracnose
Diaporthe phaseolorum	Fruit rot
Fusarium solani	Fusarium wilt
Phaeoramularia capsicicola	Leaf mould, leaf spot
Phoma destructiva	Phoma rot
Phytophthora capsici	Phytophthora blight, fruit rot
Rhizoctonia solani	Rhizoctonia
Sclerotinia sclerotiorum	Sclerotium rot, pink joint, stem canker
Pseudomonas solanacearum	Brown rot, bacterial wilt
Xanthomonas campestris pv. *vesicatoria*	Bacterial spot of fruit, stem and seedling blight
Alfalfa mosaic virus	
Cucumber mosaic virus	
Tobacco mosaic virus	

from white, yellow, green, through to shades of purple to almost black. There are also some cultivars with striped, blotched fruit and also some with a different pigmentation on their shoulders. There are many named cultivars, including F_1 hybrids. Different geographical areas tend to have their own preferences, often relating to local culinary habits and preference when selecting and developing their cultivars. By far the majority of selected material is largely devoid of spines which is common among several *Solanum* species. However, some tend to have spines on the stem, petioles, foliage, pedicels and also on calyces. The plants of some cultivars can grow to a height of 2 m. The species is a weak perennial, but normally cultivated as an annual, although at the end of the fruiting season the plants can be pruned back for re-growth. However, this is not recommended as it can result in a build up of soil-borne pests (including eelworm) and systemic pathogens, especially viruses whose symptoms and influence become more apparent in the following season.

Site and soils

The soil pH should be maintained at approximately 6.5. The crop is grown on raised beds in areas which are likely to receive heavy rainfall, otherwise it is grown on ridges with channel irrigation or on the flat. Eggplants do not succeed well in waterlogged or clayey soils where they become more prone to soilborne pathogens. The crop succeeds best on soils with high nutrient levels which have had decomposed manure incorporated during site preparation.

Plant production

Some growers propagate the plants for the following season by layering long shoots into containers of compost, but this method has the same pathogen risks as stated for carrying the plants over to the following season. In some areas of South America scions from selected plants are grafted on to root stocks of *Solanum torvum* which is a vigorous species and provides some protection from root knot eelworm. Seed is sown in prepared seedbeds containing decomposed manure or other composted material which has been incorporated in the bed's surface. The seed is thinly sown in drills 1–2 cm deep. The drill is filled in and soil surface firmed down. The drills are usually 10 cm apart. The seedbeds are maintained in a moist condition but protected from heavy rains or drying out. Some growers soak the seed in water for up to 24 h before sowing to reduce the risk of sporadic germination although good quality seed should avoid the need for this.

Crop establishment

The seedlings are planted out when approximately 10 cm high. The overall plant population per unit area should depend on the vigour of the cultivar. Ridges are 60–100 cm apart with plants 60–75 cm apart within the rows. When produced on the flat a square planting pattern can be used with the plants

50 cm each way, or in rows from 75–100 cm apart and plants within the rows 50–75 cm apart.

Cultural requirements

The newly planted areas should be kept weed free until the plants have established and are large enough to compete with weeds. A mulch is usually applied soon after planting to reduce the amount of hoeing or hand weeding required and also help to reduce the overhead irrigation requirement. However, there should always be sufficient available soil water to avoid flower bud or flower drop resulting from dry soil conditions.

A compound fertilizer N:P:K ratio of 15:15:15 applied as a side dressing at a rate of 40 g per plant. When the plants are expected to continue cropping over a long period of 4 months or more further side dressings can be applied at approximately the same rate.

Harvesting

The ripe fruit can be expected to be ready for harvest approximately 8–14 weeks from planting out, although this will very much depend on the cultivar and local conditions. The individual fruit are harvested while still firm by cutting with a sharp knife or secateurs about 2 cm above the green calyx thereby leaving a short stem on each fruit.

Pests and pathogens

The main pests and pathogens of eggplant are listed in Table 10.3.

Solanum tuberosum (Potato, Irish Potato, White Potato)

Origins and types

The alternative common names for this species distinguish it from the sweet potato (*Ipomoea batatas* (L.) Lam.) which is in the botanical family *Convolvulaceae*. *Solanum tuberosum*, which originates from South America, is increasing in popularity and importance in the subtropics and tropics.

Traditionally commercial and home garden crops of potato have been produced vegetatively from tubers, generally referred to as 'seed potatoes'. Earlier reports by the International Potato Center (CIP, 1979) indicated an interest in the production of commercial potato crops from seed, especially for developing countries. The term now widely accepted for seed of this crop is 'true seed', often referred to as 'TPS' which distinguishes it from tubers. However, there is a tendency to refer to vegetative planting material of potato

Table 10.3. The main pests and pathogens of eggplant.

Pest or pathogen	Common names
Meloidogyne spp.	Root knot eelworm
Aphis spp.	Aphids
Empoasca flavescens	Leaf hopper
Epilachna hirta	Epilachna beetle
Gargaphia solani	Eggplant lace bug
Tetranychus spp.	Spider mites
Leucinodes orbonalis	Fruit borer
Epitrix spp.	Flea beetles
Alternaria alternata	Leaf spot and fruit rot
Cochliobolus spicifer var. *melongenae*	
Colletotrichum melongena	Anthracnose
Fusarium oxysporum	Fusarium
Fusarium solani	
Phomopsis vexans	Fruit rot, Phomopsis rot
Pseudomonas solanacearum	Bacterial wilt
Rhizoctonia solani	Damping off
Sclerotinia sclerotiorum	Basal rot
Verticillium albo-atrum	Verticillium wilt
Verticillium dahliae	Verticillium wilt
Eggplant mosaic virus	
Mycoplasma-like organism	Little leaf
Leaf curl virus	

as 'seed' (i.e. the term 'seed potato' is frequently used for tubers designated as planting material).

Production of potato planting material

There are three basic systems for the production of potato planting materials. These are: (i) true potato seed (TPS); (ii) seed tubers; and (iii) vegetative propagation using *in vitro* techniques. Each of these three systems is explained as follows.

True potato seed (TPS)

Several advantages are cited for the promotion of the use of TPS which can be of significance, and with important application in developing countries. These include:

- When a supply of true seed is available all tubers produced can be used as food, thus retaining a food source and thereby securing a larger stock for food security.
- TPS saves on importation and storage of bulky tuber planting material and avoids seed tuber losses resulting from storage diseases, other storage hazards or transport and import/export delays.

- A safe supply of TPS can avoid some of the soilborne pests and pathogens which are likely to be imported on tubers for crop production. This is especially important in communities where they do not already have the potential pest or pathogen.

TPS systems can be used for the production of ware potatoes or for the production of seedling tubers for subsequent crops. Umaerus (1989) has reviewed the potential of TPS and its applications.

Tubers

The production of seed potatoes requires careful planning followed by attention to detail while the crop is in the field. National seed potato schemes usually embrace the following important points:

- Many countries have an overseeing organization which provides national guidelines, specifications and inspectorates related to seed potato production; these usually apply to the production of seed tubers which will be offered for sale. However, any producer of seed potatoes (including on-farm) would be well advised to find out about the recommended requirements and problems which are likely to occur under their local conditions.
- There is careful monitoring of the history of the site regarding previous potato and related crops including incidence of potato-related soilborne pests and pathogens.
- Provision is made to provide sufficient isolation (i.e. isolation by distance) from neighbouring crops likely to be a source of pests and pathogens with special reference to transmission of viruses by insect vectors.
- There are regular inspections of the tuber-bearing crop to remove and destroy potato plants showing scheduled virus symptoms.
- The typical tuber for planting is similar in size to a chicken egg (i.e. between 40 and 60 g); larger tubers are usually cut into pieces so that each portion contains at least one 'eye'.
- Field production of a single crop of seed potatoes in a growing season, depending on the cultivar, can result in a multiplication rate of 10:1. In recent years workers have formulated protocols for higher ratios than this, for example two systems of vegetative propagation using *in vitro* techniques are referred to below with significantly higher multiplication rates.

Vegetative propagation using in vitro *techniques*

Several propagation methods for the rapid multiplication of potato propagules have been developed and are widely adopted for the production of disease-indexed planting materials. The protocol for potatoes is described by FAO (2010). The method includes *in vitro* culture of plant material found to be disease free followed by further subculturing and production of rooted propagules which can eventually be used for the production of mini-tubers. One of the systems described

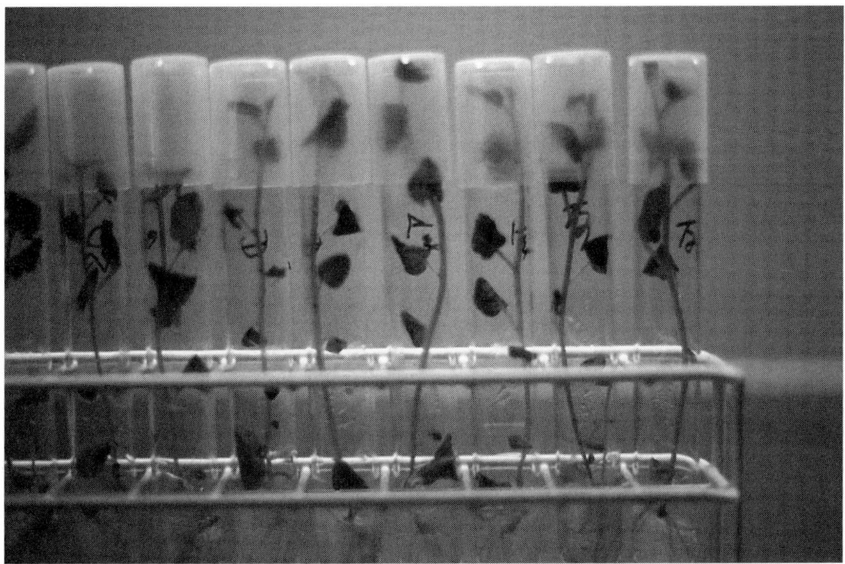

Fig. 10.6.   In vitro potato shoots cultured from a meristem tip immediately prior to subculturing into separate plantlets.

using *in vitro* techniques is capable of producing 125 *in vitro* plants from a single *in vitro* plant in 3.5 months.

A system using selected and screened tubers which uses meristem culture followed by micropropagation and final production of elite tubers for further field propagation was developed at the National Institute of Plant Pathology, Lyngby, Denmark, and has been described by George (1986). Figure 10.6 illustrates *in vitro* potato shoots which were produced from a meristem tip excised from a selected tuber found to be true to cultivar type and free of specific potato viruses and *Corynebacterium sepedonicum* (bacterial ring rot) immediately prior to subculturing. This system produces a fivefold propagule increase every 8 weeks and finally produces Nuclear Stock seed potatoes which are passed on to the National Seed Potato Multiplication Scheme for further multiplication via the Danish Institute of Potato Research.

Production of ware potatoes for food

Rotation

Proposed plots should not have produced potatoes or other solanaceous crops in the previous 3 or 4 years. Several crop rotation patterns are used where soil-borne potato pest and pathogens have been prevalent. The rotation sequence can be a cereal crop (e.g. maize) followed by a legume or a grass fallow and then potato in the fourth year. The specific crops in the rotation will obviously depend on what the subsistence farmer needs and whether or not there are any livestock to be catered for.

Site and soils

The potato crop is best suited to cool temperatures, ideally not exceeding 25°C. This crop is therefore often more successful at higher altitudes of approximately 1000 m. However, the temperature responses of available cultivars should be taken into account.

Potatoes require a soil pH of 5.3–6.0. There can be a higher incidence of potato scab (*Streptomyces* spp.) at pH levels higher than this range, especially where there is free lime in the soil. The crop benefits from soils which have a high organic content. Compacted or stony soils both tend to produce misshapen tubers. The crop succeeds best with ground-level irrigation rather than overhead irrigation because the main foliar diseases easily infect foliage where there is high relative humidity or when the foliage is frequently wet.

Crop establishment

The 'seed' tubers should ideally have been stood in wooden trays, with the eye end uppermost and allowed to produce short shoots or 'sprouts'; this process is widely referred to as 'chitting'. The 'seed' tubers should not be chitted in the dark as this promotes the production of weak and spindly shoots which are vulnerable to damage during planting and also tend to produce a poor yield.

The crop is planted on the flat and subsequently earthed up in ridges, or ridges are first made and the seed tubers planted in the top of the ridges. This latter system is frequently used when ridge-and-furrow irrigation is to be used. The tubers are planted so that the bottom of the tuber is 20 cm from the soil surface. Distances of 30 cm apart within the row and 30–60 cm between rows are used depending on the vigour of the cultivar and its predicted time to maturity; the quicker maturing cultivars ('earlies') are planted at the higher plant densities per unit area.

Cultural requirements

The crop is earthed up (ridged) as part of the hoeing for weed control thus maintaining the ridges in the case of ridge and furrows or creating ridges over the emerging plants when planted on the flat. Ridging provides the following advantages:

- There is a useful volume of friable soil in which the tubers form and grow.
- The developing stolons are encouraged to grow into the loose soil.
- The depth of soil covering the tubers helps prevent 'greening' and sun scorch of developing tubers which may otherwise be exposed.
- The depth of ridged soil gives the tubers some protection from spores of pathogens which are likely to wash down from the haulm and infect the tubers.

The amount of available water should be reasonably constant: dry soil conditions followed by wet conditions can cause misshapen tubers (often called 'dollies'), a result of re-growth of almost fully developed tubers.

Prior to harvesting

Some foliar diseases (e.g. late blight) can infect tubers as well as the haulm and foliage; the haulm can also impede the harvesting operation. To reduce these potential problems the haulm can be removed once the tubers have stopped increasing in size. There are two methods:

- Cut the haulm off just above the tops of ridges and dispose of it (if infected it should be burned) or pull the haulm up and dispose of it.
- Use desiccating chemicals which are suitable for use on larger areas, but are not normally economical for small farmers.

Harvesting

A crop produced from seed tubers is ready for harvest from approximately 3–4 months from planting, depending on the cultivar. Individual tuber-bearing plants can be harvested as required when produced as a very minor crop. But when left in the ground after they are harvestable they can be prone to increased damage from rodents. In some tropical areas monkeys can also be responsible for crop losses. The optimum stage to dig the tubers is when the tuber skins have started to set firm (i.e. just beyond the stage when the skin is loose enough to be moved by thumb pressure). Ideal conditions, guidelines and timing for lifting (harvesting) the mature tubers are:

- Lift when tuber skins start to set.
- Lift when soil conditions are relatively dry.
- Take care to avoid mechanical damage (damage from forks or other implements) during lifting.
- Separate the damaged and diseased tubers from the satisfactory tubers as lifting proceeds.
- Keep the harvested tubers out of the sun or bright light.
- Protect tubers from rain and surface water.

Postharvest

Immediately after harvesting the tubers should have a period for the skins to harden (suberize), this is sometimes referred to as 'wound healing', and reduces the possibility of water loss and deterioration during storage. Ware tubers (i.e. those stored for food supply) can be stored in sheds or specially constructed stores depending on the quantities. Fig. 10.7 shows two designs of potato stores suitable for the tropics demonstrated at Bvumbwe Research Station, Malawi.

Details for storage of tubers produced by the protocol for Quality Declared Seed of vegetatively produced crops are outlined by FAO (2010).

Pests and pathogens

The main pests and pathogens of *S. tuberosum* are listed in Table 10.4.

Fig. 10.7. Two designs of potato stores (a and b) suitable for tropical conditions.

Table 10.4. The main pests and pathogens of the potato, *S. tuberosum.*

Pest or pathogen	Common names
Leptinotarsa decemlineata	Colorado beetle
Globodera rostochiensis	Golden potato nematode
Globodera pallida	White potato-cyst nematode
Nacobbus aberrans	False root-knot nematode (mainly South America)
Phthorimaea operculella	Potato tuber worm
Ralstonia solanacearum	Bacterial wilt
Erwinia carotovora pv. *atroseptica*	Black leg
Corynebacterium sepedonicum	Bacterial ring rot
Rhizoctonia solani	Stem canker or black scurf
Spongospora subterranean	Powdery scab
Streptomyces spp.	Common scab
Potato leaf roll virus	Leaf roll
Potato virus X	Mild mosaic
Potato virus Y	Severe mosaic
Potato spindle tuber virus	Spindle virus

Further Reading

Anon. (1983) *Pest Control in Tropical Tomatoes.* Centre for Overseas Pest Research, London.

De Bokx, J.A. and van der Want, J.P.H. (eds) (1987) *Viruses of Potatoes and Seed-potato Production*, 2nd edn. Pudoc, Wageningen, The Netherlands.

Heuvelink, E. (ed.) (2005) *Tomatoes.* CAB International, Wallingford, Oxon, UK.

Kinet, J.M. and Peet, M.M. (1997) Tomato. In: Wien H.C. (ed.) *The Physiology of Vegetable Crops.* CAB International, Wallingford, Oxon, UK, pp. 207–258.

Wein, H.C. (1997) Peppers. In: Wien H.C. (ed.) *The Physiology of Vegetable Crops.* CAB International, Wallingford, Oxon, UK, pp. 259–293.

11 *Leguminosae* – Beans and Related Crops

This botanical family contains a large number of genera cultivated as vegetables, grain and fodder. Collectively the genera are frequently referred to as legumes, but can be divided into 'fodder legumes', 'grain legumes' (pulses) and 'vegetable legumes', therefore forming three groups. Some of the species can be considered as belonging to more than one of these groups, for example the French or dwarf bean (*Phaseolus vulgaris*) is cultivated as a green vegetable whose immature pods are cooked but when left to fully mature and produce dry seeds is considered to be a grain legume (or pulse crop).

Some of the fodder legumes are grown or used in rotation by subsistence farmers with other non-legume crops as part of their ongoing system of crop rotation. The choice of species often depends on the livestock species the farmer keeps. The species in the family *Leguminosae* develop root nodules in symbiosis with specific bacteria (*Rhizobia* species) which are capable of fixing atmospheric nitrogen and thereby adding to the soil's nitrogen levels.

Some of the vegetable legumes are considered to be particularly useful from an agronomic and culinary point of view when included in subsistence farmers' cropping programmes. These include hyacinth bean, Lima bean, pigeon pea, winged bean and yard long bean.

Fodder Legumes

The following fodder legumes are suitable for the tropics and subtropics:

- *Centrosema pubescens* Benth. – centro – this species is a vigorous climber, not suitable on soils which dry out quickly at the start of the dry season.
- *Leucaena leucocephala* (Lam.) de Wit. – jumbie bean, white popinac – jumbie bean is an erect semi-shrub-like plant. This species requires a well-drained alkaline soil and is suitable for the higher rainfall locations. Some farmers sow it at a wide spacing interplanted with a suitable grass fodder species.

- *Macroptilium atropurpureum* (DC) Urban. – siratro – this is a perennial with a climbing habit, capable of rooting at its nodes. It does not thrive in soils which readily dry out and it requires a weed-free environment for early establishment.
- Four species of *Stylosanthes*: (i) *S. guianensis* (Aublet) Sw. – stylo; (ii) *S. hamata* (L.) Taubert. – Caribbean stylo; (iii) *S. humilis* H.B.K. – Townsville stylo; and (iv) *S. scabra* J. Vogel – shrubby stylo. *Stylosanthes humilis* is an annual crop. The other three *Stylosanthes* species are perennial. All four species are well adapted to the higher rainfall areas although these species do not tolerate waterlogged or saline conditions. *Stylosanthes guianensis* is tolerant of relatively low soil fertility.
- *Aeschynomene americana* L. – joint vetch – this species, which originates from tropical America, is grown as a green manure in addition to being used as a forage crop.
- *Crotalaria juncea* L. – sunn hemp – this is a fibre crop but is also grown as a green manure. This species is fast growing and competes well with weeds.

Vegetable Legumes and Pulses

The vegetable species whose agronomy is discussed in this chapter are:

- *Phaseolus vulgaris* L. – dwarf bean, French bean, green bean, snap bean;
- *Phaseolus lunatus* L., syn. *P. limensis* Macf. and *P. inamoensis* L. – Lima bean, sieva bean, butter bean, Madagascar bean, Burma bean;
- *Dolichos lablab* L. syn. *D. purpureus* L., *Lablab niger* Medik. – hyacinth bean, bovanist bean, lablab bean, lubia bean;
- *Voandzeia subterranea* (L.)Thou. ex DC. syn. *Vigna subterranea* (L.) Verdc. – bambara groundnut, earth bean, ground bean, Madagascar groundnut, vouandzou;
- *Vigna unguiculata* (L.) Walp. – cowpea, yard long bean;
- *Cajanus cajan* (L.) Millsp. – pigeon pea, Congo pea, red gram, dhal, no-eye pea; and
- *Psophocarpus tetragonobolus* (L.) DC. – winged bean, asparagus bean, Goa bean.

Phaseolus vulgaris (Dwarf Bean, French Bean, Green Bean, Snap Bean)

Origin, types and uses

The species originates from Central America where several wild forms can still be found. Plant breeders in several areas of the world have worked with the species in order to produce improved cultivars. There are therefore a very large number of cultivars generally available from seed companies and other organizations. The majority are bush types but there are also climbing ('pole') types which are frequently referred to as 'climbing French beans'.

Many of the local selections in use by subsistence farmers have good pest and pathogen tolerance but are not all very high yielding. It still remains necessary to identify and use these local traits in seed improvement programmes. Some of the important characters which are available include resistance to bean common mosaic virus, bean anthracnose, halo blight and common blight.

Some of the cultivars developed towards the end of the 20th century have included 'stringless' types suitable for their immature green pods and processing. However, these are not widely distributed among subsistence farmers who still tend to use derivatives of the older cultivar 'Contender' which is very suitable for harvesting as tender green pods or for the production of dried 'haricot' beans.

Soil and nutrition

Phaseolus vulgaris tolerates slightly acid soil conditions: soils with a pH of 5.4–6.5 can be used successfully but a pH below 5.4 is likely to induce aluminium and manganese toxicities. The general N:P:K ratio applied during seedbed preparation is 1:2:2, generally at a rate of $30\,g/m^2$.

Sowing and crop establishment

The sowing depth for bush or climbing cultivars is approximately 3 cm.

Bush or dwarf cultivars are sown in rows 45–90 cm apart, with approximately 15 cm between plants in the rows.

The climbing cultivars are usually sown in double rows 60 cm apart after locally procured cane frames or supports have been erected approximately 30 cm apart; one or two seeds are sown at the foot of each cane or support. The distance between double rows is 90 cm if more than one double row is grown.

A friable mulch applied immediately after sowing is beneficial, especially for the climbing types as it is difficult to hoe within close proximity of the supporting structures.

Harvesting

The young pods are harvested as a green vegetable. Otherwise the dwarf or bush types are generally considered to be ready for a once-over harvest as dried beans when the earliest pods are dry and parchment-like and the remainder of the pods have turned yellow. The seeds' moisture content at the time of harvesting should be between 20–25%; seed maturity is confirmed by opening sample pods, the seeds should be fully developed with a mealy texture. Under good growing conditions the flowers tend to continue setting until relatively late in the season especially if the green pods are picked as they are ready.

The plants are cut and placed in windrows for further drying prior to hand separation of the beans from the dry pods or before threshing; small threshing machines can be very useful for large quantities, especially in community and cooperative ventures. The entire threshing operation should be organized so as to ensure both the minimum loss from shattering and the least possible mechanical damage to the seeds which are especially susceptible to cotyledon cracking. When using a mechanical thresher a drum speed of 250–350 rpm with a concave clearing of *c.*12–20 cm should be adopted and the seeds' moisture content should not be too low otherwise excessive mechanical damage occurs during the process.

The climbing cultivars mature over a longer period than the bush types therefore they are harvested by hand as required for fresh use, or on three or more successive occasions as the older pods mature on the plants. Alternatively the plants are pulled out and dried off in windrows before seed separation as described above for the bush types.

Wilson and McDonald (1992) have evaluated six systems for the threshing of *P. vulgaris* and found that the highest quality seed was produced by open flail threshing or hand shelling. They concluded that, where feasible, manual threshing methods are superior to mechanical methods for small seed lots. Manual threshing would be expected to minimize cotyledon cracking and this is an important consideration if growers are planning to use some of their own 'on-farm produced seed' to produce their next crop; damaged seeds are also more likely to succumb to storage pests and pathogens.

Pests and pathogens

The main pests and pathogens of *Phaseolus* species are listed in Table 11.1.

Table 11.1. The main pests and pathogens of *Phaseolus* species.

Pest or pathogen	Common names
Alternaria alternata	Leaf blight
Ascochyta boltshauseri, Ascochyta phaseolorum and other *Ascochyta* species	Ascochyta leaf spots, pea blight
Aspergillus sp.	'Baldheads' and 'snake-heads' of seedlings
Cercospora canescens	Brown blotch
Colletotrichum lindemuthianum	Anthracnose
Fusarium oxysporum f. sp. *phaseoli*	Yellows, wilt
Fusarium solani f. sp. *phaseoli*	Root rot
Phaeoisariopsis griseola	Angular leaf spot
Pleospora herbarum	Red nose, leaf spot
Rhizoctonia solani	Damping off, stem canker
Sclerotinia sclerotiorum	Sclerotinia wilt, stem rot, watery soft rot, white mould
Pseudomonas syringae	Bacterial brown spot
Pseudomonas syringae pv. *phaseolicola*	Halo blight, grease spot
Xanthomonas campestris pv. *phaseoli*	Common bacterial blight, fuscous blight
Bean common mosaic virus	

Phaseolus lunatus syn. *P. limensis* and *P. inamoensis* (Lima Bean, Sieva Bean, Butter Bean, Madagascar Bean, Burma Bean)

Origin and types

The Lima bean is believed to have its centre of origin in Guatemala, South America and subsequently spread northwards to Mexico and southern USA; it also spread further southwards in South America. It was later introduced into Asia and Africa. This species is an important crop for subsistence farmers in dry tropical areas.

There are two main annual cultivar types: (i) small annual bush; and (ii) climber. Some local types have a high level of hydrocyanic acid which is released from an enzyme present in the seeds, but it is broken down on cooking. The bush types have a shorter growing season than the climbing types; but overall the production of dried Lima bean seed requires a longer season than dried French beans.

Several cultivars bred in the USA have proved to be useful in Africa, especially those with eelworm resistance. However, it is important that improved cultivars developed in cooler climates are able to produce sufficient viable pollen in higher temperatures in order to produce satisfactory yields.

Soil and nutrition

Soils described for *P. vulgaris* are suitable but with a pH ranging from 6.0 to 6.5 as the Lima bean is more sensitive to acid soils than *P. vulgaris*. Ideally, the climbing Lima beans benefit from soils with higher moisture availability as they have a longer growing season. The crop responds to a satisfactory soil phosphorus level and fertilizers or other materials known to enhance phosphorus availability should be added during site preparation if there is any evidence of low levels of this element. The crop also responds to top dressings of a nitrogenous fertilizer at the rate of 40 g/m^2 especially when leaching is likely to have occurred.

Lima beans should be rotated with unrelated crops. Some farmers grow it as a companion crop to maize or sorghum. In some areas it follows a yam crop, thus utilizing the remaining support system after the yams have been harvested.

Sowing and crop establishment

Lima beans are usually grown as a supported, climbing crop by subsistence farmers or as a ground crop, especially when cultivated for the production of dried pulses.

Sowing and plant population densities are similar to *P. vulgaris*.

Harvesting

The bush types can usually be expected to produce their first harvest approximately 3 months from sowing and be productive over a period of 5–6 weeks.

The climbing types can be ready for harvest in 4 months from sowing and continue cropping for a further month or so.

Pests and pathogens

Pests and pathogens are the same as listed in Table 11.1 for *Phaseolus* species with the following additions:

- *Meloidogyne* spp. – root knot eelworm;
- *Corynebacterium flaccumfaciens* – bacterial disease;
- *Phytophthora phaseoli* – downy mildew; and
- *Diaporthe phaseolorum* – pod blight.

Dolichos lablab syn. *D. purpureus, Lablab niger* (Hyacinth Bean, Bovanist Bean, Lablab Bean, Lubia Bean)

Origin, types and uses

The origin of the hyacinth bean is thought to be Asia, with secondary centres of diversity in parts of tropical Africa. It is well established as a food crop in India and South-east Asia, and is also widely grown by small farmers in Africa, being an important subsistence farmer crop in many countries especially the Sudan. There are local selections and local cultivars, but the species is largely cross-pollinated and there is some confusion between 'local varieties' and cultivars. According to Purseglove (1974) there are two botanical varieties, one with longer pods and the long axis of the seed parallel to the suture, the other with shorter pods and the long axis of the seed at right angles to the suture. There are several well-established cultivars in cultivation.

The hyacinth bean is an herbaceous perennial, although more frequently grown as an annual, it is cultivated as either a pulse or for the young pods which are used as a fresh vegetable; some farmers, especially in India, use it for stock feed. It is especially suited to dry environments and is sown on river banks following a fall in water levels in addition to other relatively dry sites. It is usually supported on canes or other structures when cultivated as a vegetable, or on the flat when grown for stock feed or pulses.

Soil and nutrition

Ideally this crop requires a well-drained and fairly fertile soil with a pH of 5.5–6.0. The crop responds to available phosphatic fertilizers. A balanced NPK general fertilizer can be applied during site preparation and as a top dressing after plant establishment, but before the start of anthesis. Each of these applications are usually at the rate of $40\,g/m^2$.

Table 11.2. Pests and pathogens of *D. lablab.*

Pest or pathogen	Common names
Agromyza sp.	Stem flies
Adisura atkinsoni	Pod borer
Aphis fabae	Black bean aphid
Cerotoma ruficornis	Bean leaf beetle
Exelastis atomosa	Red gram plume moth
Meloidogyne spp.	Root knot eelworm
Cercospora canescens	Leaf spot
Cercospora dolichi	Leaf spot
Elsinoe dolichi	Scab
Leveillula taurica	Powdery mildew
Macrophomina phaseolina	Root wilt, stem blight
Uromyces phaseoli	Rust rot
Xanthomonas phaseoli	Bacterial blight

Sowing and crop establishment

The plants require a support system which should be erected before sowing.

Seed is either sown in ridges or beds in rows 60 cm apart and depending on potential seed germination at a distance such that plants are established 30 cm apart within the rows.

The plants should be kept weed free while becoming established. Some irrigation is required in arid areas.

Harvesting

It should be emphasized that some types and cultivars require cooking before being eaten.

The crop takes approximately 3–4 months before mature seeds have formed. Some growers regard the plant as a short-term biennial and keep the plants growing in order to produce a later, second crop.

Pests and pathogens

The pests and pathogens of *D. lablab* include those listed in Table 11.2.

Voandzeia subterranea syn. *Vigna subterranea* (Bambara Groundnut, Earth Bean, Ground Bean, Madagascar Groundnut, Vouandzou)

Origin and uses

This species is cultivated in West Africa where it is believed to have originated. It was subsequently introduced to parts of South-east Asia, especially the Philippines and Indonesia, and also limited areas of South America, particularly in Brazil and Surinam.

It is an important crop for subsistence farmers, especially on poor soils in areas with high temperatures and low rainfall. Many workers in the field have regarded it as an under-utilized crop with important potential for improving food security and deplore the fact that there has been very little research or development input into this crop.

Bambara is cultivated for its edible seeds which are usually consumed before they ripen, although the ripened seeds are also ground to flour for culinary preparations. The seeds are produced under the ground as a result of the recurving inflorescences pushing into the soil after pollination.

Morphological types

There are no named cultivars cited in the literature, although local selections are maintained by farmers in some areas. The types which are generally considered more suited to cultivation are those which have less of a trailing habit and form plants which are more clumped. Hepper (1970) considers that there are two botanical varieties in cultivation: (i) var. *spontanea* which includes wild forms; and (ii) var. *subterranea* which includes the cultivated forms.

Selections or local cultivars are based on internode length, plant height, plant form and spread, flower colour (yellow, green or orange), number of seeds per pod, pod colour and size.

Soil and nutrition

The crop is suited to light soils which facilitate harvesting. Seedbeds should be prepared to form a fine tilth. This is especially important as the nuts are formed following flower fertilization on the long recurving flower shoots which enter the soil. The optimum pH is in the range of 5.5–6.5 although the crop requires adequate available calcium for satisfactory development of the nuts. The optimum N:P:K fertilizer ratio is 1:2:2 applied during site preparation ideally at a rate of $40\,g/m^2$. The crop responds to bulky organic manures although these also may not be available to the poorer farmers to whom this crop is especially important. When bulky organic manures are incorporated during site preparation care should be taken to reduce the nitrogen in base dressings otherwise there will be excessive foliar growth at the expense of nut yield.

Sowing and crop establishment

The unshelled seed can be sown, although it is generally considered preferable to shell before sowing. Shelled seed is sown approximately 7 cm deep in flat-topped ridges at stations 50–60 cm apart with 50 cm between the ridges. It is especially important to use a ridged system where waterlogging is likely to occur otherwise a flat seedbed may be used with the same sowing density, provided there is a fine tilth. The plants are earthed up during anthesis which commences approximately 30–50 days from sowing. The earthing-up operation also includes weed control; an earlier weeding may be necessary before anthesis if weeds become established soon after sowing.

Harvesting

The crop should be lifted from the soil as soon as it is ready, otherwise the seeds may germinate before harvesting, especially in areas where the rainy season commences immediately after crop maturity. The maturity of the subterranean pods should be determined by spot examinations from the time after the aerial foliage commences to turn yellow. Under dry conditions the leaves will also wilt and wither as the seeds reach maturity.

The mature seeds are either ploughed out or hand dug, depending on the scale of operation. The lifted plants with pods are left in windrows on the soil surface to further dry for up to 3–5 days.

The seed pods are hand threshed, although in cooperative or group production the seeds can be extracted with a groundnut thresher.

Pests and pathogens

Pests and pathogens of bambara groundnut include *Meloidogyne* spp. (the root knot eelworm) and several viruses have been identified in farmers' fields in Nigeria (see Further Reading list).

Vigna unguiculata (Cowpea, Yard Long Bean)

Origin, types and uses

The primary diversity centre of *V. unguiculata* is thought to be Africa and the main types cultivated in Africa, especially West Africa, are thought to be of African origin. India has been recognized as a secondary centre from where the types grown in Asia have been developed. Local lines and cultivars have become well established in both regions. In addition to these places of origin the cowpea is also an important crop in South America.

There are dwarf (determinate and semi-determinate) and climbing (indeterminate) types. Some authorities also use the 'eye patterns', which describes the shape and eye colour of the pigmented area surrounding the seed's hilum, of individual cultivars and land races in cultivar descriptions.

There are many named cultivars and locally named selections, some have resistance to root knot eelworm and cultivars have also been developed for arid environments.

The long green pods are used as a fresh green vegetable and the maturing seeds are used before they ripen. The seed pods may also be left to dry for the production of the pulse crop. There is also a short pod type which is grown as a climber. The leaves, immature pods, young and developing seeds, in addition to the mature and ripe seeds of all types are used for culinary purposes. Cowpea is a very popular crop with women farmers in parts of Africa. These farmers use part of the early harvestable leaves and immature pods for their own domestic requirements and subsequently trade the dried seed for cash. The remaining plant material provides a very useful crop of hay for livestock. Some subsistence farmers intercrop cowpea with a grain crop such as maize, sorghum or local millet types, while others prefer a single cowpea crop, especially if planning to sell the grain to market traders.

Soil and nutrition

The crop succeeds better on light soils, such as sandy loams, ideally with a high organic content. Weak or slow growth can be stimulated by top dressing with a nitrogenous fertilizer such as ammonium sulfate, at a rate of 30 g/m^2, but excessive available nitrogen will produce vegetative growth at the expense of inflorescence and pod development.

Sowing and crop establishment

The crop is usually grown on the flat, but on ridges in situations where irrigation will be required. *Striga gesnerioides* (witch weed) is a parasitic weed that can be a serious problem in this crop.

Dwarf types are sown in rows 45 cm apart with 15 cm between plants within the rows.

The climbing types should have the supports in place before the seed is sown; the plant can reach a height of 2 m. Seeds are sown in rows 75 cm apart and 30 cm apart within the rows. Figure 11.1 illustrates climbing yard long beans in Indonesia.

Harvesting

Picking for use as a green vegetable can be done as the pods are considered to be sufficiently developed in length but still tender. The young leaves are also used as a cooked vegetable and can often be lightly harvested from approximately 3 weeks following seedling emergence.

Crops scheduled for dry bean production should be harvested in stages when the pods start to dry and turn brown; if left too long after this stage the pods will shatter and seed will be lost. Some cultivars and selections have a strong tendency to shatter. As with the processing of all dry bean seed, care must be taken not to damage the seeds during their extraction or mechanized processing. The cowpea plant material provides a very useful fodder.

Fig. 11.1.   Yard long beans growing on supports. The pods shown are at the picking stage as green vegetables.

Pests and pathogens

The main pests and pathogens of *V. unguiculata* are listed in Table 11.3.

Table 11.3.   Pests, pathogens and parasitic weed of *V. unguiculata*.

Pest or pathogen	Common names
Bruchidius atrolineatus	Pea and bean weevils
Callosobruchus spp.	Pea and bean weevils
Meloidogyne spp.	Root knot eelworm
Striga gesnerioides	Witch weed, (a parasitic weed)
	Pulse storage pests and pathogens

Cajanus cajan (Pigeon Pea, Congo Pea, Red Gram, Dhal, No-eye Pea)

Origin, types and uses

The main centre of diversity is tropical Africa with a secondary centre in India. The crop is especially important in India. ICRISAT, the institute for breeding and development of pigeon pea, is at Hyderabad, India. Pigeon pea is an important crop for large-scale production and processing, especially in the West Indies. It is also recognized as an important crop for subsistence farmers and is a very useful crop for soil improvement. Pigeon pea is very suitable for the dry tropics but not very satisfactory in the humid tropics where it is frequently attacked by a range of insect pests. The crop requires relatively dry weather during anthesis to enable successful insect pollination to take place.

Immature pods may be picked for cooking otherwise the main crop is used as a dried pulse in a range of food preparations.

Soil and nutrition

This species produces reasonably well on most soil types but does not tolerate waterlogging. It has a deeply penetrating root and can therefore survive drought conditions. Fertile soils which are well drained are ideal to produce the crop. The optimum soil pH is 5.5–6.5.

There should be a sufficient break between successive crops when Fusarium wilt has occurred.

The application of nitrogen is not normally considered necessary either as a base or subsequent top application but general fertilizers containing both phosphorus and potassium can be usefully applied during site preparation.

Sowing and crop establishment

Seed is sown on ridges or raised beds. The rows are 50–180 cm apart (depending on the vigour of the cultivar) and the seeds are sown two to three per station at 60 cm intervals within the rows. The young seedlings are subsequently thinned to one per station following their emergence.

Harvesting

The mature seed can be harvested 4–6 months from sowing, depending on the cultivar and local climatic conditions. Harvesting and drying are the same as described for cowpea. Early ripening pods should be hand harvested to avoid loss by shattering before the main cutting and seed separation is done. Pigeon pea is less prone to mechanical damage than many other legume pulses; small

Table 11.4.   Pests and pathogens of *C. cajan*.

Pest or pathogen	Common names
Meloidogyne spp.	Root knot eelworm
Ancylostomia stercorea	Pod borer
Aphis spp.	Aphids
Empoasca facialis	Green leaf hopper
Exelastis atomosa	Red gram plume moth
Heliothis armigera	Bollworm, corn eelworm, gram caterpillar
Colletotrichum lindemuthianum	Anthracnose
Diplodia cajani	Stem canker
Fusarium oxysporum f. sp. *phaseoli*	Fusarium root rot
Phoma cajani	Collar rot, stem canker
Physalospora cajanae	Collar rot
Phytophthora cajani	Stem blight
Uredo cajani	Rust
Uromyces phaseoli	Red rust
Xanthomonas cajani	Stem canker
Mosaic virus	Mosaic

threshing machines with a low drum speed are useful in cooperative ventures otherwise flailing is usually the common method of seed extraction.

Postharvest pruning

Some cultivars can produce a second crop, either following directly on or in the following year. A light pruning of vegetative growth can stimulate new growth for production of a following crop, although this is not advisable if serious pests or pathogens have occurred in the plants.

Pests and pathogens

The pests and pathogens of *C. cajan* are listed in Table 11.4.

Psophocarpus tetragonobolus (Winged Bean, Asparagus Bean, Goa Bean)

Origin, types and uses

This species is thought to have originated in tropical Asia, and was subsequently introduced into the Caribbean and parts of West Africa.

The crop is a well-established smallholder and home garden vegetable in the more humid areas of South-east Asia, including Papua New Guinea and Indonesia. It has been identified as potentially suitable for wider adoption in the tropics and some temperate areas, especially for subsistence farmers. Named cultivars are only of local significance although there are distinct character differences between local types.

The plant provides a range of culinary uses: (i) the immature pods are used as a green vegetable when about half developed; (ii) the leaves can be used as a leafy green vegetable; (iii) the tuberous roots can be eaten; and (iv) the mature seeds may be consumed following specific preparation to render them safe and palatable for eating.

The winged bean plant is a climbing perennial, although it is mainly cultivated as an annual. Many farmers grow it as a climber with vertical support to a height of up to 2 m, as illustrated in Fig. 11.2.

Fig. 11.2. Winged bean plant growing on wire mesh support in the Philippines.

Table 11.5. Pests and pathogens of *P. tetragonobolus*.

Pest or pathogen	Common names
Meloidogyne spp.	Root knot eelworm
Aphis craccivora	Groundnut aphid, cowpea aphid
Maruca testulalis	Spotted pod borer, leaf miner
Synchytrium psophocarpi	Leaf spot
Psophocarpus ringspot virus	

Crop production

There is little available information on the crop's nutrient requirements, but it responds to bulky organic manures used in soil preparation. A base nutrient dressing incorporated in the seedbed consisting of N:P:K 1:6:4 at a rate of $40\,g/m^2$ would take into account its response to a low nitrogen: phosphorus ratio, although a higher level of nitrogen would be required when there is significant leaching.

Seed is sown in rows 1–1.5 m apart with a final distance of 50–60 cm between plants within the rows.

Harvesting

The young and immature pods are harvested while still tender, before the pod reaches its full length. The leaves may also be picked and cooked as a green, leafy vegetable, but ideally not during the peak flowering and pod development stage.

Pests and pathogens

The main pests and pathogens of winged bean are listed in Table 11.5.

Further Reading

Davis, J.H.C. (1997) *Phaseolus* beans. In: Wien H.C. (ed.) *The Physiology of Vegetable Crops.* CAB International, Wallingford, Oxon, UK, pp. 409–478.

Erskine, W., Muehlbauer, F. and Sharma, B. (eds) (2009) *The Lentil: Botany, Production and Uses.* CAB International, Wallingford, Oxon, UK.

Smartt, J. (1976) *Tropical Pulses.* Longman Group Ltd, London.

Thottappilly, G. and Rossel, H.W. (1997) Identification and characterization of viruses infecting bambara groundnut (Vigna subterranea) in Nigeria. International Journal of Pest Management 43(3), 177–185.

12 Leafy Vegetables

A wide range of tropical species are consumed as leafy vegetables in the tropics, especially by subsistence farmers and their dependents. Many are collected in the wild or from appropriate weeds which have been identified as edible and are retained as a food source. In some cases, such as *Gynandropsis gynandra* (spider flower), selections have been made and seed made available so as to produce it as a reliable and more uniform and stable indigenous crop.

The botanical families and the genera dealt with in this chapter are considered to be the more important leafy vegetables cultivated by subsistence farmers; some other species in which the culinary use of the leaves is secondary to the main morphological part of the plant are mentioned under individual crop species described elsewhere in this volume.

Many perennial shrubs and trees growing in the wild or planted for shelter or other purposes can provide leafy culinary materials. In many areas they provide the only leafy food materials throughout the dry season. Many of the minor leafy species are considered to be 'pot herbs', 'leafy green vegetables' or simply as 'greens' or 'herbs'. Appendix 1 in this volume lists indigenous species which are locally important, and are either gathered from the wild or cultivated as land races or local varieties. Others as described in this chapter are more important especially as defined by the amounts cultivated. These are from the following plant families: *Asteraceae, Amaranthaceae, Basellaceae, Chenopodiaceae, Malvaceae, Portulacaceae.*

Asteraceae (Formerly *Compositae*)

Lactuca sativa L. (lettuce)

Origin and types
Lettuce has its centre of origin in the Mediterranean region where it is believed to be derived from *Lactuca serriola* L. which remains as an indigenous weed

species. *Lactuca sativa* var. *asparagina* is believed to have originated in China. There is a very wide range of variation within the cultivated forms of *L. sativa* which are divided into four main types. The divisions, based on morphological characters are:

1. *Lactuca sativa* var. *capitata* L. – the cabbage or head lettuce which is generally subdivided into crispheads (iceberg) and butterheads. The butterheads have relatively soft-textured leaves with a greasy appearance. As in the crispheads the leaves of mature plants form a heart.
2. *Lactuca sativa* var. *longifolia* Lam. – the cos or romaine lettuce. In this group the upright relatively narrow crisp-textured leaves form a closed head.
3. *Lactuca sativa* var. *crispa* L. – the 'leaf' or curled lettuce does not form a heart but has a loose head of leaves. Some of the cultivars of this type have fringed leaves.
4. *Lactuca sativa* var. *asparagina* Bailey (syn. var. *angustana* Irish) – all the forms of this group have typical fleshy stems which are the main culinary attraction, especially in Asia where the group may be referred to as 'stem lettuce' or 'asparagus lettuce'. A typical cultivar is 'Celtuce'. Some of the members of this group have a light-grey leaf colour.

In the first three types of lettuce there is a wide range of cultivars with green leaves but cultivars with various degrees of red pigmentation are also available.

Lettuce is an important commercial salad crop in the temperate regions and is also grown at higher altitudes in the tropics, mainly as a market crop, especially in Asia. It is not considered to be a significant crop for subsistence farmers although it may be useful as a pot herb or as a boiled vegetable in some areas, but eating it raw, as in temperate regions, is frequently considered to be a health risk in tropical areas.

Site, soils and pH

Lettuce is generally considered to be a crop for the higher elevations of approximately 1000 m. Ideally the crop requires a well-drained light loam which has a high decomposed organic content. The pH should be approximately 6.5; if significantly lower a liming material should be added during preparation and this will also ensure that there will not be a calcium deficiency which can otherwise cause some physiological problems. A balanced N:P:K 7:7:7 fertilizer should be raked in the seedbed or planting area at the rate of 40 g/m^2.

Crop establishment

The plants are raised from seed, either sowing direct or planted out from a seedbed or raised in modules. The seed frequently becomes thermo-dormant in soil temperatures above about 25°C and either fails to germinate or germination can be sporadic. Figure 12.1 demonstrates uneven emergence of a lettuce seed sample when germinated at a temperature of 26°C. It is therefore often an advantage to produce the transplants in shaded beds, the optimum germination temperature for most cultivars is 18–21°C. A thin layer of mulch over the bed immediately after sowing will help to avoid high soil temperatures.

The crop is sown direct, or planted out on the flat in rows 30 cm apart; if the crop is grown with furrow irrigation, the ridges are 1 m apart. The optimum

Fig.12.1. Lettuce seed sample displaying effect of thermo-dormancy resulting in uneven germination in high soil temperatures.

distance between plants within the rows in either system depends on the morphology of the cultivar; the cos types are generally less spreading than the butterheads, but plant densities which are too high per unit area will add to the risk of the fungus 'grey mould' (*Botrytis cinerea*) resulting in crop losses. The transplants should be planted so that the finished soil surface is slightly below each plant's initial seedling leaves. The crop raised from direct seeding is usually thinned in stages of approximately 10 days, until the optimum inter-row spacing for the cultivar has been achieved. The thinnings provide a useful food source, if not used in this way they should not be left within the plot, otherwise they will become a focal point for *Botrytis*, especially under humid or wet conditions.

Harvesting
The lettuce heads are cut close to the soil surface; the lower stem is trimmed to remove yellow or off-colour leaves and the cut head is then kept shaded, top down, immediately following cutting until taken for domestic use. The cut heads wilt very easily, therefore it is important to always keep them out of direct sun and in a cool place until required.

Pests and pathogens
The pests and pathogens of *L. sativa* are listed in Table 12.1.

Lactuca indica L. (Indian lettuce, sawi rana, kuban kayu rana)

This perennial *Lactuca* species is believed to have originated in China where it can still be found in the wild in the more tropical areas. The leaves are used as a

Table 12.1.   Pests and pathogens of *L. sativa*.

Pest or pathogen	Common names
Meloidogyne spp.	Root knot eelworm
Botrytis cinerea	Grey mould, storage rot
Bremia lactucae	Downy mildew
Cercospora lactucae	Cercospora leaf spot
Erwinia spp.	Soft rot
Marssonina panattoniana	Ring spot, anthracnose
Pleospora herbarum f. sp. *lactucum*	Leaf spot
Pythium spp.	Damping off
Sclerotinia sclerotiorum	Drop, watery soft rot
Sclerotium rolfsii	Southern blight
Septoria lactucae	Leaf spot
Pseudomonas cichorii	Leaf blight
Rhizoctonia solani	Root rot
Arabic mosaic virus	
Lettuce mosaic virus	
Lettuce yellow mosaic	
Tobacco ringspot virus	
Tomato black ring virus (ringspot strain)	

cooked green vegetable and may also be used to wrap small fish before frying. It is cultivated in China, India, Malaysia, Indonesia and the Philippines.

Plant production and crop establishment
The most widely used method of plant production of *L. indica* is from seed although it has been reported to also be propagated from root cuttings (Tindall, 1983). *Lactuca indica* also regenerates from the root stocks following the dying down of the aerial stems, forming a weak ratoon crop.

Seed is usually sown in a seedbed and the seedlings subsequently transplanted when large enough to handle at a spacing of approximately 30 cm square. It is necessary to keep the planted area weed free until the plants form a rosette of leaves on the soil surface.

Harvesting
Lactuca indica grows to a height of 120 cm. Leaves which arise from the main stem are harvested from the base of the plant upwards, as required for culinary purposes.

Amaranthaceae

This family contains several species which are important leafy vegetables including:

- *Amaranthus cruentus* (L.) Sauer – African spinach, bayam, bush greens, Chinese spinach;

- *Amaranthus dubius* C. Martius ex Thell.; and
- *Amaranthus tricolor* L.

There are also some *Amaranthus* species grown as grain crops for the production of flour which is used for making unleavened bread or mixed with wheat flour for the production of leavened bread, especially in South America. The most important species for grain production are: *Amaranthus hypochondriacus* L., *Amaranthus caudatus* L. (Inca wheat) and to a lesser extent *A. cruentus*. The grain amaranths are subject to lodging when grown in the humid tropics and also lodge very easily in high soil nitrogen regimes.

Celosia argentia L. (cock's comb, green or white soko, quail grass, Lagos spinach) has been regarded as an under-utilized crop by Badra (1993) who cites five other *Celosia* species as being of minor importance as a source of food: (i) *Celosia trigyna*; (ii) *Celosia globerosa*; (iii) *Celosia insertii*; (iv) *Celosia leptostachya*; and (v) *Celosia pseudovirgat*. There are both green and so-called white forms of *C. argentia*.

Amaranthus and *Celosia* species are considered very important leafy vegetables for subsistence farmers and their dependents, these crops are usually included in home garden schemes.

Vegetable amaranths and *Celosia*

Origins and types

The centres of diversity for *Amaranthus* and *Celosia* are in Central and South America, India and South-east Asia with secondary centres in Africa. These genera are annuals.

There is a wide range of morphological types in different areas of the world, varying in height from approximately 25 cm to 2 m, with a range of branching habits. There is also a range of leaf types including those with red pigmentation and bicolour, and leaf shape ranging from thin and smooth to rugose. The morphological types adopted vary from one geographic area to another depending on local preferences. There are cultivars and lines which have some resistance to *Choanephora cucurbitarium* (leaf and stem wet rot) which can result in significant crop loss.

Soil and pH

The optimum soil pH is 6.5, a higher pH tends to cause iron deficiency, indicated as lime-induced chlorosis; these genera do not tolerate acid conditions. The crops are very widely grown throughout the tropics on a range of soil types although ideally they grow best on loamy, well-drained soils. The amaranths can survive periods of low water availability better than most other leafy vegetable genera.

Crop establishment

The seed of both *Amaranthus* and *Celosia* is relatively small (their 1000 seed weights are approximately 0.3 and 1.0 g, respectively). The crop is grown from

Table 12.2. Pests and pathogens of *Amaranthus* spp. and *Celosia* spp.

Leafy vegetable	Pest or pathogen	Common name
Amaranthus spp.	*Hymenia recurvalis*	Leaf caterpillar
	Prodenia litura	Leaf caterpillar
	Alternaria amaranthi	Blight
	Choanephora cucurbitarium	Leaf and stem wet rot
	Pythium spp.	Damping off
	Rhizoctonia spp.	Damping off
Celosia spp.	Lilac ring mottle virus	
	Spinach latent virus	

seed sown thinly on the surface of a prepared seedbed, covered with a sprinkling of fine soil or sand, lightly firmed with a board or plank. The seedlings are subsequently planted out when large enough to be handled (approximately 4 weeks from sowing), unless they have been sown direct; most growers prefer production of transplants in seedbeds. Whichever method is used the seedlings are thinned out soon after emergence. As most subsistence farmers harvest the leaves as required for culinary purposes, the plants' final spacing is approximately 50 × 50 cm, although this will depend on the cultivar's potential vigour (i.e. height and spread of the local selection or cultivar used).

In some areas where the taller and more apically dominant types are grown for leaf production each plant is 'stopped' (i.e. the growing point is pinched out) to produce a more bushy plant.

Top dressings of a nitrogenous fertilizer such as ammonium sulfate can be applied after crop establishment at a rate of 30 g/m^2, especially if growth is slow. An organic equivalent can be used but it should be well decomposed to minimize the risk of ammonia scorch.

Harvesting

Generally, the leaves are harvested as required. In some parts of Asia the stems of the taller types are peeled and cooked. The majority of growers pick off leaves as required as a pot herb although early harvesting from young plants should be light.

Pests and pathogens

The pests and pathogens of *Amaranthus* spp. and *Celosia* spp. are listed in Table 12.2.

Basellaceae

There are two important genera in this family: (i) *Ullucus*; and (ii) *Basella*. *Ullucus tuberosus* Caldas (ulluco) is discussed in Chapter 14 of this volume 'Andean Tubers and Roots and Crops of the *Lamiacae* and *Apiaceae*'.

Basella alba L. syn. *B. rubra* L. (Indian spinach, Malabar spinach, vine spinach)

Origins and types
This vegetable is a perennial species. It is thought to have originated in Asia and has become well established as a leafy vegetable in China and India from where it has spread to other parts of Asia and Africa. The leaves and tips of new shoots are picked and cooked as a pot herb. There are two leaf pigmentation types, red or green, and local selections have been made from each.

Soils
The crop is not very demanding regarding its root environment, although it generally thrives better on a well-drained site.

Crop establishment
Basella leaf crops are produced on the flat, supported on poles up to 250 cm high or on a trellis which has been erected horizontally at a suitable height for harvesting. Indian spinach is propagated either from seed or 15 cm-long stem cuttings; ideally the vegetative cuttings should be taken from selected plants and inserted direct into their final quarters. The seedlings are raised in a seed-bed and planted out when large enough to handle. The bed will usually require some temporary shade or protection from the wind when in exposed situations.

Plants grown on the flat are usually planted 45 cm apart in rows 70 cm apart. When planted along and under prepared trellis structures the planting distances are approximately 120 × 120 cm. Some growers make use of boundary or internal fences and hedges to support the crop. Post-planting application of a mulching material will assist in control of weeds in addition to reducing water loss. Irrigation is required in dry periods to ensure a continued supply of young shoots and a top dressing of a nitrogenous fertilizer at a rate of 40 g/m^2 can be applied immediately prior to irrigation when foliage growth is weak or slow.

Flowering shoots should be removed before they become too dominant. They are not of any culinary value although interestingly the fruit juice has been recorded as having been used as an ink.

Harvesting
Although widely referred to as 'spinach', it is the terminal shoots as well as the larger leaves which are gathered. Edible crops are usually available approximately 1 month from planting cuttings; plants raised from seed can usually be harvested from approximately 2 months from sowing. Care should be taken not to damage the main stem when harvesting. According to *Sturtevant's Edible Plants of the World* (Hedrick, 1972) the leaves have been infused to make a tea in India.

Pests and pathogens
The main pest of *Basella* is *Meloidogyne* spp. (root knot eelworm).

Chenopodiaceae

The main genera in this family cultivated as vegetables are:

- *Beta vulgaris* L. subsp. *esculenta* – beetroot, red beet;
- *B. vulgaris* subsp. *cycla* Moq. – chard, Swiss chard, spinach beet, leaf beet;
- *Spinacia oleracea* L. – spinach (European spinach); and
- *Chenopodium quinoa* Willd. – quinoa.

Beetroot is a popular temperate region root crop and can be grown at higher elevations of the tropics but is not considered as an important crop here for the subsistence farmer. European spinach can also be produced at higher elevations but as with beetroot is not considered in this volume.

Beta vulgaris subsp. *cycla* (chard, Swiss chard, spinach beet, leaf beet)

Origins and types
This subspecies is closely related to sugarbeet and fodder beet (mangel). It is thought to have originated in the Mediterranean and has been gradually selected for its edible leaves with their prominent rugose petioles and wide leaf blades; at the same time it has spread to many other regions from where it has been adopted as an important leafy vegetable, for example in the warmer parts of Nepal.

Chard is a biennial which requires a low temperature for vernalization, otherwise it can continue as a weak perennial. However, some of the local selections derived from indigenous types can produce flowering heads in their first year when cultivated in parts of Asia. Cultivars with green, red or yellow leaf blades and laminae are available. The cultivar 'Fordhook Giant', with a green petiole and prominent white lamina, originally bred in the USA, is now widely grown in the tropics and subtropics.

Soil and pH
The satisfactory soil pH range is 6.0–6.8. This crop is susceptible to boron deficiency, but this can be corrected by the addition of 'borax' (sodium tetraborate) at a rate of $2\,\text{g/m}^2$. Where the deficiency has been positively identified in previous crops the use of boronated fertilizers can be used, although these are not always available to the subsistence farmer due to cost or isolation from suppliers. Chard crops well in light loams or silts which have a satisfactory water supply during dry periods.

Crop establishment
The so-called 'seed' of the *Beta* species is a multigerm fruit (i.e. the so-called 'seed' contains several seeds). Plant breeders have produced monogerm 'seeds' of some of the *Beta* subspecies, namely sugarbeet and red beet. 'Seed' is sown thinly in rows 50 cm apart. The emerging seedlings should be thinned to one per station as soon as they can be handled. The young plants are thinned to a final spacing in the rows of approximately 30 cm as they develop.

Light top dressings of a nitrogenous fertilizer at a rate of 30 g/m^2 are beneficial if leaf growth is weak or there has been heavy leaching.

Harvesting

The leaves are harvested from the outside of the plant as each leaf reaches its approximate maximum size, but before any signs of deterioration show. Yellowing or decaying foliage should be removed when first seen. Leaves are detached from the plant by cutting the lamina as close to its base as possible with a sharp knife; alternatively a lamina can be held and twisted to the side to detach it from the mother plant.

The harvested leaves should be immediately taken out of the sun and kept cool prior to culinary use.

Pests and pathogens

The main pests and pathogens of chard are listed in Table 12.3.

Chenopodium quinoa (quinoa, Inca rice)

This species has its centre of origin in Peru and also occurs in Chile and the Bolivian Andes. It is widely cultivated in South America, particularly in the Andes, where it is a staple grain crop and also cultivated for its edible leaves. There is a range of morphological types described by Flemming and Galway (1995) with variations in mature plant heights, branching and non-branching, plant pigmentation at maturity and seed colour.

It is considered as an important species for food security and possibly as a potential crop for cultivation at higher altitudes in other parts of the world such as the Northern areas of Pakistan and India.

Site and soil

Quinoa is traditionally rotated with *S. tuberosum*. Some farmers sow it in bands within their maize crops.

This species is well adapted to harsh conditions and because of its initial tap root which subsequently branches the plant can tolerate soils with low available

Table 12.3. The main pests and pathogens of chard.

Pest or pathogen	Common names
Ditylenchus dipsaci	Eelworm canker
Albugo candida	White rust
Cercospora beticola	Leaf spot
Erysiphe betae	Powdery mildew
Erwinia carotovora	Bacterial soft rot
Peronospora farinosa f. sp. *betae*	Downy mildew
Pleospora betae	Black leg, damping off, leaf spot

water. Some types are capable of tolerating salt soil conditions. The crop's pH tolerance is wide, ranging from 4.8 to 8.5; this wide tolerance is a reflection of the range of ecotypes in South America although it is best suited to the light loams. All these factors make the species suitable for cropping by subsistence farmers although it is a labour-intensive crop. The crop responds favourably to light dressings of nitrogenous fertilizers at a rate of $40\,g/m^2$ applied as a base dressing, however, excessive available nitrogen will result in plants lodging without increase in their seed yield.

Crop establishment

The seed is sown approximately 2 cm deep in moist soil, with the rows 30 cm apart. A wider spacing may be used for the taller land races. The seed germinates within 48 h in moist conditions. Weeds should be controlled at an early stage as the quinoa seedlings are slow in developing. In some areas the crop is earthed up while hoeing for weed control.

Harvesting

The goose-foot shaped leaves are picked while young for cooking as required and while still green. Harvesting for seed extraction is indicated when the maturing seed offers very slight resistance to the 'thumb-nail' test. This is a test to determine the stage of seed maturity before harvesting; a seed sample is taken in the palm of the hand and pressed with the thumb nail of the other hand, ripe seed will offer resistance, especially when the thumb nail attempts to cut into the seed.

As the plants approach maturity, the leaves tend to fall, and the seed heads are hand harvested once the seed hardness test has confirmed the stage of maturity. It is essential that the seed crop does not get wet once it is mature as this will result in premature germination. Further drying may be required before threshing. The crops produced by subsistence farmers are usually hand threshed on a tarpaulin or smooth, clean surface and subsequently winnowed.

The harvested seed has a natural saponin in the pericarp (seedcoat) which due to its bitterness, has acted as a deterrent to birds prior to harvest. Although

Table 12.4. The main pests and pathogens of *Quinoa*.

Pest or pathogen	Common names
Melanotrichus sp.	Quinoa plant bug
Spodoptera exigua	Beet army worm
Ascochyta hyalospora	Leaf spot
Botrytis cinerea	Grey mould
Fusarium sp.	Fusarium wilt
Peronospora farinosa	Downy mildew
Phoma exigua var. *foveata*	Leaf spot, brown stalk rot
Pseudomonas spp.	Bacterial blight
Rhizoctonia sp.	Damping off
Sclerotium rolfsii	Seed rot
Pythium zingiberum	Damping off

some of the saponin may have been removed by rain or irrigation while the crop was standing in the field it is essential that it be completely removed prior to culinary use or grinding for flour. This is either done by soaking in water or by a mechanical process. However, this is not done for seed that will be retained for production of the next crop.

The plant remains are used for fodder or animal bedding after the seed heads (panicles) have been gathered.

Pests and pathogens

The main pests and pathogens of *Quinoa* are listed in Table 12.4.

Malvaceae

Hibiscus esculentus L. syn. *Abelmoschus esculentus* (L.) Moench. (okra, lady's finger, gombo)

Origins and distribution

Okra is thought to have originated in the Ethiopian region of Africa but is now widely cultivated in Africa, especially in the Sudan, Egypt and Nigeria; it is also very important in many other tropical areas including Asia, Central and South America.

Okra is cultivated for: (i) its leaves which are used as a cooked green vegetable; (ii) the immature and semi-mature edible beaked capsules, usually referred to as 'pods' which are also used in this way; and (iii) the mature capsules which are dried and stored in parts of Africa for local use in the high temperature season. The mature fruit which is a beaked capsule can have a high mucilaginous content, depending on cultivar, which is dried and stored and used to advantage in the dry season for the preparation of soups and stews.

There are many cultivars and local selections, many with very specific characteristics such as capsule length and shape, spiny or smooth capsules, ripe seed colour, partial pigmentation of foliage and level of mucilaginous content (especially important for those which will be stored as the dried capsule). The more modern cultivars produced in the USA, such as 'Clemson Spineless' produce fruit relatively early on more compact plants than the traditional local African types. Figure 12.2 illustrates a typical fruiting okra plant.

Soil and pH

The crop produces reasonably well on most soil types although a fertile loam is preferable. Cultivars which are susceptible to root knot eelworm do not produce a satisfactory crop on infected light or sandy soils. This species tolerates soils which are slightly acid and will grow successfully in soils with a pH of 6.0–6.8.

Crop production

Farmers often interplant okra with tomatoes or peppers, alternatively those who are producing cassava, yams or some other tropical tubers may choose to cultivate okra as a possible intercrop with these tuberous species.

Fig. 12.2.   An okra plant in flower and bearing edible capsules. Photograph courtesy of The Real IPM Company (K) Ltd (www.realipm.com).

Some writers suggest that the seed is recalcitrant although the author has not found sound evidence of this in practice or in the literature. Many growers soak the seeds overnight in water prior to sowing to overcome the hardness of the seed's testa.

Harvesting and storage

The younger leaves and young fruit are gathered as required, although this should be kept to a minimum if a larger proportion of the potential crop is intended for either semi-mature or dried capsules. However, plants can be kept in production over a long period by continually harvesting the immature capsules. Seed maturity is indicated by the capsules becoming brown or grey, depending on the cultivar. The capsules ripen from the base of the plant upwards and are liable to shatter with loss of seed. The harvesting of ripening

Table 12.5. The main pests and pathogens of okra.

Pest or pathogen	Common names
Bemisia spp.	Whitefly
Heliothis spp.	Caterpillar
	Leaf beetles
Meloidogyne spp.	Root knot eelworm
Oidium abelmoschi	Powdery mildew
Erysiphe cichoracearum	Powdery mildew
Verticillium sp.	Verticillium wilt
Fusarium sp.	Fusarium wilt
Ascochyta abelmoschi	Ascochyta blight, pod spot
Botrytis sp.	Stem and capsule disease
Choanephora cucurbitarium	Okra fruit rot, leaf wet rot
Fusarium oxysporum f. sp. *vasinfectum*	Fusarium wilt
Fusarium solani	Fusarium foot rot
Glomerella cingulata	Anthracnose fruit rot
Pythium spp.	Damping off
Rhizoctonia solani	Damping off
Okra leaf curl virus	
Yellow vein virus	
Okra mosaic virus	

capsules should be done in sequence from the base of the plant upwards rather than as a single harvest; this is also essential for the collection of 'on-farm' produced seed.

The dried capsules are either stored whole or ground to form a fibrous powder for food storage; in some areas the seeds alone are ground for food.

Pests and pathogens
The main pests and pathogens of okra are listed in Table 12.5.

Portulacaceae

Talinum triangulare syn. *Talinum racemosum* (L.) Rohrb. (water leaf, grass bologi, Philippine spinach, Surinam spinach, gbure)

This member of the *Portulacaceae* is the species most widely cultivated rather than being collected from the wild as with the other *Portulacaceae* species listed in Appendix 1. *Talinum triangulare* originated in South America and was subsequently introduced into Central Africa; it is widely cultivated in parts of Asia and the Caribbean. There are local selections and named cultivars. The modern ornamental cultivars of this species are bred from another species and are not considered to be edible.

Water leaf can grow up to a height of approximately 90 cm although if grown in short days it will only grow to an approximate height of 50 cm. It can be used as a fresh stock feed in addition to being consumed as a green vegetable.

Soil

Talinum triangulare reaches its best cropping potential in soils which are well drained with a high organic and nutrient content, although it can crop fairly well on poorer soils. It is often cultivated in beds on river banks.

Crop production

Some subsistence farmers rely on self-sown seed for a continuous supply of the crop. However, the crop is usually produced in raised beds and is established either from seed or vegetative cuttings.

When the plants are produced from seed the seedlings are usually produced in a seedbed or raised in modules. The young plants are transferred to their final positions when approximately 5–8 cm high.

When the plants are produced from cuttings 10–15 cm long stem cuttings are taken from healthy, relatively vigorous mother plants. After removing about a third of the leaves from the base of the cutting they are inserted into a prepared cutting bed to just below the remaining lower leaf. The inserted cuttings usually require shading and also frequent fine overhead watering at this early stage, and the cuttings root in approximately 4 weeks.

Plants produced from seed or cuttings are planted out in their final positions when 5–8 cm high when produced from seed, or when rooted if propagated from cuttings. The plants are grown on in rows 30 cm apart with approximately 25 cm between plants within the rows.

The young plants should be kept weed free in their early stages. If growth is slow or weak a top dressing of a soluble nitrogenous fertilizer such as ammonium sulfate can be applied at a rate of $40 \, \text{g/m}^2$. Plants which are well maintained and kept in a vegetative state rather than being allowed to flower and seed early can be expected to crop for up to 12 months. Some growers remove the tip of the shoot (often referred to as 'stopping') to encourage the production of lateral shoots soon after the young plants have established in their final positions.

Harvesting

When the foliage is eaten it can leave a slightly bitter taste on the palate as a result of the low level of oxalic acid present (thought to be from 1 to 2%).

At the first harvest terminal shoots which have not been stopped, but were allowed to fully develop, are cut when approximately 15–20 cm long. Similar length lateral shoots are taken at the subsequent harvests or at the initial harvest if the plants have not been 'stopped' earlier.

Pests and pathogens

No serious pests or pathogens are listed in the literature.

Further Reading

Leaf vegetables, general

Food and Agriculture Organization (FAO) (1991) *Gyandropsis gyandra* (L.) Briq. *A Tropical Leafy Vegetable Its Cultivation and Utilization*. FAO Plant Production and Protection Paper 107. FAO, Rome, Italy.

Martin, F.W. and Rubert, E. (1979) *Edible Leaves of the Tropics*, 2nd edn. Agricultural Research Service, Southern Region, United States Department of Agriculture (USDA), Mayaguez, Puerto Rico.

Oomen, H.A.P.C. and Grubben, G.J.H. (1978) *Tropical Leaf Vegetables in Human Nutrition*, 2nd edn. Joint edition, Royal Tropical Institute and Amsterdam and Orphan Publishing Company, Willemstad, Curaco.

Quinoa

Risi, J.C. and Galwey, N.W. (1994) The *Chenopodium*, grains of the Andes: Inca crops for modern agriculture. *Advances in Applied Biology* 10, 145–216.

Simmonds, N.W. (1984) Quinoa and relatives. In: Simmonds, N.W. (ed.) *Evolution of Crop Plants*. Longman Group, Harlow, UK, pp. 29–30.

Amaranthus

Williams, J.T. and Brenner, D. (1995) Grain amaranth (*Amaranthus* spp.). In: Williams, J.T. (ed.) *Cereals and Pseudocereals*. Chapman and Hall, New York, pp. 129–186.

13 Araceae, Convolvulaceae, Dioscoreaceae, Euphorbiaceae – Tropical Tubers

This chapter covers vegetables in the families *Araceae*, *Convolvulaceae*, *Dioscoraceae* and *Euphorbiaceae*, commonly referred to as tropical tubers.

Araceae

There are two genera of major importance as crops in the *Araceae*, these are *Colocasia* and *Xanthosoma*.

There are several other genera which are of local importance. Lebot (2009) describes other members of the *Araceae* which are of economic importance. The *Colocasia* species have a peltate leaf form which distinguishes them from *Xanthosoma* species which have sagittate leaves.

Colocasia species

There are two important *Colocasia* species but a mix of common names is used in some areas and texts for these two species. Therefore to ensure clarity they are:

- *Colocasia esculenta* var. *esculenta* (L.) Schott – referred to as taro; and
- *C. esculenta* var. *antiquorumi* (Schott) Hubbard and Rehder – referred to as eddoe.

Cropping area, suitable climate and environment

Eddoe and taro are cultivated in a wide range of tropical areas. Taro, which is an important crop in West Africa and the West Indies, is also a staple food crop in rainforest areas of the Pacific Islands. The vegetable species of *Colocasia* can be grown successfully a few degrees further south or north of the Tropics than many of the other cultivated genera in the family *Dioscoreaceae* and therefore have a wider global distribution. Both crops tolerate a wide range of tropical climatic conditions, although eddoe tolerates low temperatures and drought more than taro.

These two crops produce well in deep organic or loamy soils which have satisfactory drainage rather than the heavy clayey or sandy soil types; supplementary irrigation is necessary during dry periods. The optimum soil pH for both crops is pH 5.5–6.5. The eddoes tolerate waterlogged conditions, while taro will produce well under irrigation in drier or higher altitudes.

Planting material and planting

Both of these species are vegetatively propagated. Ideally propagules should be obtained from selected plants which are pest and pathogen free, however, some viruses are symptomless and can only be detected in plant material by specific laboratory tests. Seed is rarely produced and normally only used in breeding programmes.

The vegetative propagation methods for each species are:

- Eddoe – lateral corms (cormels) which are produced towards the top of the main corm. The number of propagules can be increased if required by cutting cormels into several pieces, each piece having part of the corm's crown.
- Taro – the top of the main root, often referred to as the 'crown' or 'headset'. A few of the taro species produce stolons, or runners, which are also used as propagules.

A protocol for the production of improved quality *Colocasia* planting materials is described by FAO (2010). The propagules of both crops are planted out in furrows 80 cm apart and 40 cm within the rows. The furrows are filled in after planting and earthed up; frequent earthing up contributes to tuber formation and weed control. Side dressings of an inorganic fertilizer (N:P:K 6:8:8) are advantageous immediately prior to the first earthing up after shoots emerge. This is applied along the rows at the rate of 30 g/m^2.

Harvesting

Immature leaves and petioles can be harvested from time to time as a leaf crop if required. Although the leaves are acrid resulting from calcium oxalate crystals, the acridity is reduced with boiling in water. Immature tubers may also be harvested. However, the tuber crop takes approximately 6–9 months from planting to reach maturity prior to satisfactory storage, by which stage the foliage has started to turn yellow; earlier stages of tuber maturity do not store well although they may be harvested for immediate culinary use before reaching maturity. Figure 13.1 illustrates mature taro tubers on sale in Fiji.

Pests and pathogens

The pests and pathogens of *Colocasia* (taro) are listed in Table 13.1.

Fig. 13.1. Mature taro tubers on sale in a Fiji market.

Table 13.1. The pests and pathogens of *Colocasia* (taro) species.

Pest or pathogen	Common names
Tarophagus proserpina	Taro leaf hopper
Papuana spp.	Taro beetles
Meloidogyne spp.	Root knot eelworm
Phytophthora colocasiae	Taro leaf blight
Colocasia bobone disease virus (CBDV)[a]	
Taro bacilliform virus (TaBV)[a]	
Pythium spp.	

[a] Alomae is a disease believed to be caused by infection with both CBDV and TaBV (see Revill *et al.*, 2005).

Xanthosoma sagittifolium (L.) Schott. (tannia, tanier, yautia, cocoyam)

In the following text this species is referred to as tannia.

Origins and types

The genus *Xanthosoma* originates in the tropics of South America where, in addition to the West Indies, it is still an important food crop. It was introduced by missionaries into West Africa where it was given the common name of cocoyam because of its resemblance to *Colocasia* although tannia has sagittate leaves.

There are obvious morphological differences between clonal selections and cultivars of tannia, especially with regard to leaf vein and petiole pigmentation, corm shape, size, external colour and internal flesh colour. Tannia tends to grow better in drier situations than the *Colocasia* species.

Planting material and planting

Tannia is a perennial but it is cultivated as an annual from vegetativly propagated material; either the small cormels or the propagules are cut from the main corm. The prepared material is planted as described for taro and cultivated in the same way. A protocol for the production of improved quality *X. sagittifolium* planting materials is described by FAO (2010).

Harvesting

The young leaves may be harvested as a green leaf crop although care has to be taken because of the presence of oxalic acid. The tubers are usually harvested after the foliage has started to turn yellow. The duration of tuber development varies from approximately 6 months to 1 year according to where in the tropics it is cultivated; presumably this is also due to locally preferred cultivars or selections interacting with local environments. The mature crop can usually be left *in situ*; in some areas it can be simply pulled out when harvesting, although care should be taken to gather all the corms and cormels, otherwise the crop is better lifted with suitable hand tools.

Pests and pathogens

The pests and pathogens of tannia are given in Table 13.2.

Table 13.2. Pests and pathogens of tannia.

Pest or pathogen	Common names
Ligrus ebenus	Tannia beetle
Araecus fasciculatus	Storage beetle
Pratylenchus spp.	Root lesion eelworm
Meloidogyne spp.	Root knot eelworm
Pseudomonas solanacearum	Bacterial soft rot
Xanthomomas campestris	Bacterial leaf spot
Erwinia carotovora pv. *atroseptica*	Tuber soft rot
Phytophora spp.	Leaf blight
Fusarium oxysporum	Dry rot
Rhizoctonia sp.	Rhizoctonia root rot
Sclerotium rolfsii	Sclerotium root rot
Dasheen mosaic virus (DsMV) (a potyvirus)	

Convolvulaceae

Ipomoea batatas (L.) (sweet potato)

Origins, types and uses

The sweet potato is sometimes given the English common name of yam in some areas which can be misleading. It is likely that this arises from the African dialect name *nyami*, referring to edible roots of *Dioscorea*.

The sweet potato originates from north-western areas of South America. The cultivated types subsequently spread to the Pacific Islands from where they reached Asia. Sweet potatoes are currently produced as a major food crop throughout Africa, Asia and the Americas; it is an important staple in the highland areas of Papua New Guinea. Although this species is a perennial it is normally cultivated as an annual crop from cuttings.

There are some 16 species and natural hybrids in the *Batatas* section of *Ipomoea* described by McDonald and Austin (1990). The taxonomy and botany of sweet potatoes are discussed by Lebot (2009). Named cultivars are available and plant breeders continue to produce improved cultivars. Some stocks maintained by subsistence farmers are morphologically variable as a result of seedling off-types which have been allowed to remain in growers' crops and have produced tubers.

This species is cultivated for its edible root tubers which are usually boiled or baked; the growing tips of the aerial shoots and leaves ('vines') are used as a cooked or parboiled green vegetable. Tubers from large-scale commercial crops are processed for production of starch and alcohol. The vines are also used for stock feed.

Climate and environment

The crop thrives on the lighter sandy soils with irrigation. The crop does not tolerate waterlogging, although it responds well to irrigation in dry situations or seasons. In the wetter or high rainfall areas it is grown on ridges. Ideally, the temperature should be at least 24°C and although it grows at 10°C and above, cropping is poor at the lower temperatures. The crop is not very satisfactory in shade.

Soil and site preparation

Sweet potatoes are frequently regarded as a basic food crop for resource poor farmers although it also responds well to improved cultural conditions and organic manures. However, materials with a high potential nitrogen release should be used sparingly as high nitrogen tends to delay tuber production and also leads to unnecessary vigorous vine and foliage growth.

The optimum soil pH for the crop is between 5.6 and 6.6. It is sensitive to soils low in available calcium, boron or magnesium. A pre-planting general base dressing containing a balanced N:P:K fertilizer should be applied when obtainable at a rate of $40\,g/m^2$. Where soils are known to be low in potassium a top dressing containing this element should be applied approximately halfway between planting and the estimated harvest date.

A satisfactory rotation of 4 or more years is essential in order to reduce the risk of soil-transmitted pests and pathogens. When the crop follows a legume nitrogen can be withheld.

Planting material

Plant production from true seed is only used in breeding programmes. Vegetative propagation is commonly used for production of replacement planting stock including:

- Apical stem cuttings – this is the most widely used method and is achieved by taking apical stem cuttings approximately 28 cm long from maturing plants and removing the leaves from the lower half of each cutting. Each cutting is inserted at an angle in the prepared soil. Crop succession can be achieved by taking cuttings from a recently established planting, and planting these cuttings in turn so that a further crop can be established to ensure continuity of supply.
- Sprouts derived from tubers – selected small tubers are planted just below the surface in soil or sand beds which are kept damp by overhead watering. Each developed shoot is severed from its tuber when approximately 25 cm high. The severed shoot cuttings are planted direct into prepared final positions as described below.

A protocol for the production of improved quality sweet potato planting material is described by FAO (2010).

Cultivation

The crop is traditionally grown on mounds or ridges which are up to 70 cm apart, with up to 30 cm between plants in the rows of ridges. If the mound system is used, the plant density overall is about the same except that three cuttings are usually inserted on each mound. The ridges or mounds are earthed up as the crop becomes established, it is generally considered that earthing up reduces the incidence of weevil damage. Mounds are used on sites with a high water table.

Early weed emergence is controlled by hoeing although once the crop produces its vines which cover the soil surface weeds are smothered and cannot compete. In this way sweet potato is often regarded in rotations as a 'weed-smothering' crop.

Harvesting

Subsistence farmers usually harvest the tubers as needed. The tubers reach marketable size in 4–5 months from planting; although time to harvest can be more or less, depending on the cultivar and overall environmental conditions. Growers should harvest the crop when it is at optimum maturity so as to minimize crop loss from rats and/or sweet potato weevil. When necessary, plants which have to remain in the soil until required for household use should again be earthed up to reduce 'greening' of the tubers and also potential weevil damage. Irrigation should be stopped approximately 2 weeks prior to estimated harvest date. Great care should be taken when lifting the crop to avoid mechanical damage or bruising of tubers which can predispose them to storage diseases. Freshly harvested tubers should be protected from direct sunlight when lifted to minimize the risk of sunscald and 'greening'. In the cooler areas, freshly harvested tubers which are be stored should have a short period of 'wound healing' at approximately 30°C with the optimum relative humidity at 80–90%.

Storage

Tubers intended for storage should be very carefully selected. Cultivars which have low dry matter contents have a higher respiration rate and a shorter shelf life (Lebot, 2009). Material which is mechanically damaged, showing weevil damage or any evidence of disease should not be included with the stored crop. The requirements for the maintenance of stored tubers have been described by Stathers *et al.* (2005).

There are three main storage methods:

- In a building or storeroom – either packed in baskets or other suitable containers. Buildings and structures which are to be used for storage should be cleaned and fumigated in readiness to receive the tubers.
- Sand- or grass-lined trenches – the tubers are placed in the trench and then covered at ground level with dried grass (while ensuring provision of air ventilation) held in position by a layer of soil. It is customary to erect a shelter over the storage site to minimize sun and rain damage.
- Clamps in shady areas usually built up at ground level or on low mounds (depending on flooding risk) – the clamps are provided with air ventilators and are enclosed with cut grass or other disease-free plant materials, a final soil layer is then put over the grass material.

There are many local methods of preservation, including slicing, drying or parboiling followed by drying which extend the product's shelf life.

Pests and pathogens

The pests and pathogens of *Ipomoea* species are listed in Table 13.3.

Ipomoea aquatica Forsk., syn. *I. reptans* Poir (kang kong, water convolvulus, Chinese water spinach, swamp cabbage)

This crop is commercially produced in China and other parts of Asia but it is also an important source of nutrients for subsistence farmers. There are two botanical forms: (i) the 'land form'; and (ii) the 'semi-aquatic' form. The young vegetative shoots of both types are cooked as a green vegetable, parboiled or eaten raw. The plant also occurs in the wild where it is gathered as a fodder crop.

Cultivation

The land form is mainly cultivated in Indonesia and Malaysia and is an herbaceous perennial that requires relatively moist conditions. It is produced from seed sown on raised beds and there are several reputable seed companies in Asia who market the seed. If the seed has been produced and stored on-farm, germination

Table 13.3. Pests and pathogens of *I. batatas* (sweet potato) also *I. aquatica* (kang kong).

Pest or pathogen	Common names
Meloidogyne spp.	Root knot eelworm
Cyclas spp.	Sweet potato weevils
Euscepes postfasciatus	West Indian sweet potato weevil
Dickeya dadantii	Bacterial root and stem rot
Pseudomonas solanacearum	Bacterial wilt
Streptomyces ipomoea	Soil rot
Sweet potato feathery mottle virus (SPFMV)	
Sweet potato chlorotic stunt virus (SPCStV)	
SPFMV and SPCStV (dual infection)	Sweet potato virus disease (SPVD)
Sweet potato latent virus (SPLV)	
Sweet potato chlorotic fleck virus (SPCFV)	
Sweet potato virus G (SPVG)	
Sweet potato leaf curl virus (SPLCV)	

may be improved by soaking the seed for approximately 24 h. A finely divided organic mulch applied to the seedbeds after sowing will assist moisture retention and suppress early weed emergence, especially with slow seedling emergence. The young plants are either thinned to approximately 15×15 cm or transplanted into rows at 15 cm within the row and up to 45 cm apart. The land form of the crop requires frequent irrigation, but does not require running water like the semi-aquatic form.

The semi-aquatic form is an aquatic herbaceous perennial, although relatively short lived due to frequent propagation by cuttings to replace harvested or dwindling stocks. It is cultivated in South-east Asia and parts of Africa. Subsistence farmers take multiple harvests although many commercial producers take a once-over harvest and replant.

Beds for the semi-aquatic form tend to regenerate from propagules derived from floating shoots in established beds, although to enhance the quality of the stock it is preferable to regenerate by terminal cuttings of approximately six nodes taken from selected plants. Half the length of each cutting is inserted in the substrate. New beds are started on moist sites which are suitable for forming into flowing water beds immediately after the cuttings are planted; the prepared beds are similar in physical condition to rice beds at planting time. The plant density is higher than the land form outlined above: plants are established in rows approximately 10 cm apart and up to 6 cm within the rows, and the crop very soon forms a dense vegetative bed. It is important that the water source is not contaminated by disease-causing organisms. The beds are gradually flooded after planting with the water level increased to a final depth of approximately 15 cm as the plants establish.

Harvesting

The aquatic crop can be harvested after about 1 month; this encourages further aerial shoot development and further cropping.

Dioscoreaceae

Yams

The term yam is only used in this volume for members of the *Dioscoreaceae*, but it should be noted that some workers also refer to sweet potatoes and some of the tropical aroids as yams.

Origins and types

The *Dioscoreaceae* is a monocotyledonous family and the species grown for food are of tropical origin where they continue to be widely cultivated. They are an important food crop in West Africa, parts of South-east Asia, the Caribbean and many other tropical areas. There are two sections of the genus:

- Combilium which twines in a clockwise mode (to the left) and includes *Dioscorea esculenta.*
- Enantiophyllum which twines anticlockwise and includes *Dioscorea alata*, *Dioscorea rotundata* and *Dioscorea cayenensis.*

The two most important species are *D. rotundata* (originating in West Africa) and *D. alata* (originating in Asia). Other important species are *D. cayenensis* (originating in West Africa) and *Dioscorea trifida* (originating in South America) which together form the most important yield of food yams totalling at least 80% by weight of global production (FAO, 2010). Purseglove (1972) provides a botanical key for the identification of the principal edible yams and also illustrated descriptions of some 11 edible yam species including their origins and distribution. There is a tendency for some *Dioscorea* cultivation areas to give them regional, local or common names which in some cases makes it difficult to know which species or even clone is being referred to.

Suitable climate and environment

The genus *Dioscorea* generally thrives better where there is adequate rainfall or water supply. The satisfactory temperature range is 20–30°C; with the exception of *Dioscorea japonica* and *Dioscorea opposita* the cultivated species cannot survive frosts (Coursey, 1967).

Traditionally, yams are the initial crop planted after clearance where shifting cultivation or bush fallow is practised; this is a reflection of the crop's success in soils with high nutrient and organic levels. According to Purseglove (1972) the minimum annual rainfall requirement is 1000 mm, although the crop does not tolerate waterlogging.

Some agronomists in the field speculate that the cultivation of *Dioscorea* will decline in parts of Africa in favour of cassava because of yam's labour requirement for preparation of 'mounds' and traditional aerial support systems.

Soils

Yams thrive in deep loam; lighter sandy soils tend not to hold sufficient water reserves for this crop. The soil pH is not critical although soils with a pH below 5.5 should ideally have a dressing of lime.

Propagation and production of planting material

Farmers select tubers from the previous crop and either replant the small ones without division or divide the larger tubers into pieces. The tops (crowns) can also be cut from large tubers (in parts of Africa known as 'pricking') and planted for regeneration. Some species produce seed but this source of new planting material is confined to plant breeders. A protocol for the production of improved quality planting materials of yams is described by FAO (2010).

Cultivation

The prepared propagules are usually planted on ridges or mounds. Any compost or manure mixtures incorporated during site preparation should be well decomposed to avoid infection of the plantlets which are planted approximately 10 cm deep; if planted too near the surface there is a risk of them drying out before the formation of shoots. In some dry areas farmers 'cap' the planted propagules after planting with a layer of dried vegetation, such as grass. The vegetative material is put on top of the planted area and then 'capped' with a shallow covering of soil; this prevents excessive drying out, although a routine lookout for various pests, including termites, should be maintained.

Yams are normally grown up supports which keeps their 'vines' off the ground and provides room for other crops such as cucurbits, watermelons or legumes which cover the soil surface and assist with weed control. The support systems vary from the occasional tree left during land clearance to a systematic layout of vertical poles with strings radiating from the top of each pole to train individual yam plants. Some farmers use tree branches as the main vertical support with smaller supports such as palm leaves at an angle around which provide access to the main support. In areas where yams follow *Zea mays* (maize) or *Pennisetum glaucum* (Guinea corn, bulrush millet, Indian or pearl millet), the remaining crop stems are bent over at right angles approximately 60–90 cm above soil level to form a rustic horizontal framework to support the yam vines. These training systems provide the yams with better facilities for photosynthesis than if they were entwined at ground level.

As soon as the aerial parts start to show above soil level the crop can benefit from a side dressing of a general balanced inorganic fertilizer at a rate of 40 g/m^2. Nwinyi and Enwezor (1985) have reported results of fertilizer trials with *D. rotundata*.

Harvesting and storage

The yam crop is harvestable when the aerial parts start to senesce, as displayed by yellowing of the vines and foliage. The yams generally have a good storage potential provided they are free of damage, most of which is inflicted by yam beetle or received during weed control and lifting. A yam tuber's storage life starts to decline when it starts to produce shoots. As an alternative, tubers can be left in the ground. Storage is important from the point of view of food security and also for securing planting material for the next generation of the crop. The main objective is to ensure that the tubers are kept dry and in the dark. Many farmers construct a yam storage building ('barn'), although the simplest system for subsistence farmers is to gather the yams in heaps on the ground under a

Table 13.4. The pests and pathogens of *Dioscorea* species.

Pest or pathogen	Common names
	Leaf miners
	Cutworms and surface caterpillars
Meloidogyne spp.	Root knot eelworm, root galls
Crioceris livida	Leaf eating beetle
Aspidiella hartii	Scale insect
Heteroligus meles	Yam beetle
Scutellonema bradys	Yam nematode
Pratylenchus spp.	Root lesion nematodes
Cercospora carbonacea	Leaf spot
Goplana dioscoreae	Yam rust
Albugo ipomoeae-panduratae	White rust
Plenodomus destructans	Foot rot
Sclerotium rolfsii	Sclerotial wilt

form of shelter to protect them from sunlight and water. More sophisticated structures have the fundamental concept of keeping the tubers in layers off the ground with a protective roof made from thatched palm leaves or similar natural materials. Stored tubers should be inspected frequently to remove diseased tubers and also deal with any pest attacks. Attention should also be given to the control of rats and other rodents during tuber storage.

Pests and pathogens
The pests and pathogens of *Dioscorea* species are listed in Table 13.4.

Euphorbiaceae

Manihot esculenta Crantz. (cassava, mandioca, yucca)

Cassava is the common crop name used in this volume for *M. esculenta*.

Manihot esculenta is the only vegetable crop of importance in this family and is a valuable source of starchy food in the tropics. Rawel and Kroll (2003) describe it as a source of basic dietary energy for more than 500 million people in Africa, Asia and Latin America. The species is a short-lived perennial shrub; although it flowers and produces viable seed this is normally only used in plant breeding programmes. Planting material is produced from stem cuttings.

Cassava originates in South America where there is evidence of it having been a cultivated crop for several thousand years; the species was later introduced into Africa via Portuguese islands and also taken to Asia. It is a short-day species: tuber formation is reduced in a day length exceeding approximately 10h.

Cultivars and types
There are two main types: (i) 'bitter'; and (ii) 'sweet'. The difference is detected by the amount of hydrocyanic acid (HCN) contained in the parts of the plant

consumed. The two types are thought to have been derived from two different species but according to Purseglove (1974) the distinction is not justified as the two merge into each other and the toxicity of a clone varies from place to place. However, Purseglove also states that for practical purposes it is usual to divide the cultivars into:

1. Bitter cassavas with high HCN content with HCN generally distributed throughout tubers, including the core.
2. Sweet cassavas with low HCN content with HCN confined to the pheloderm of tubers.

It is interesting to note that the bitter types generally escape locust attacks and are also less prone to loss from some species of feral and wild animals. A review of this crop with special reference to the considerations on the removal of its cyanogenic potential has been described by Rawel and Kroll (2003).

Soil requirements

Well-drained sandy loams are ideal especially if they have a satisfactory organic content. Cassava can tolerate heavier clays providing the drainage is effective and the crop is planted on ridges. The crop tolerates a wide range of pH from 5.5 to 8.0. Satisfactory crops can be obtained on poorer soils, and although it will respond favourably to potassium fertilizers, excess nitrogen causes stem and leaf development at the expense of tubers. Some of the clonal selections are drought resistant. Cassava is usually intercropped with other vegetable species by subsistence farmers.

Propagation and planting

Stem cuttings of approximately 20–25 cm are taken from the more mature stems of a standing crop, each cutting with a straight basal cut and a sloping top cut. Care should be taken to avoid taking propagation materials from plants showing virus or other transmittable pests or pathogens. Some insect pests of cassava, including mites, scale insects and mealy buds may be present on mother plant stems; care must be taken to avoid transferring them to new plantings. It should be noted that whitefly (*Bemisia tabaci*) and some other insects are vectors of the crop-reducing viruses. The cuttings are inserted in the prepared planting area which has been ridged or mounded; mounds up to 60 cm are used in swampy or riverside alluvial soils. In some parts of West Africa farmers choose to insert the prepared cuttings at an oblique angle, this preference may be a reflection of using longer cuttings but other aspects of cutting orientation are discussed by Onwueme (1978). The crop can also be produced on the flat, but only where the drainage is efficient. When grown as a row crop the cuttings are inserted at 60 × 60 cm or 140 × 140 cm; higher plant densities are used on poorer soils. The crop is kept weed free in its initial stages and this is usually achieved by earthing up during the crop's establishment.

A protocol for the production of improved quality cassava planting material is described by FAO (2010).

Table 13.5. Pests and pathogens of cassava.

Pest or pathogen	Common names
	Feral pigs, rats and some other mammals including baboons and hippopotami
Phennacoccus manihot	Mealy bug
Mononychellus tanajoa	Cassava green mite
Tetranychus telarius	Red spider mite
Aonidomytilus albus	Scale insect
Saissetia miranda	Scale insect
Phytophthora spp.	Root and stem rots
Phomes lignosus	Tuber rot, white thread
Xanthomonas axonopodis pv. *manihotis*	Bacterial blight
Unspecified virus	Cassava mosaic disease (CMD)
Unspecified virus	Cassava brown streak disease (CBSD)

Harvesting and storage

Cassava produces a cluster of up to ten root tubers close to the stem base. Time from planting to harvest varies with cultivar or clonal selection, and can be from 12 to 18 months from planting. The tubers do not store well after lifting and are therefore left in the ground by subsistence farmers until required. A short piece of the woody material joining the tuber to the mother plant is left attached to the harvested tuber when storage is contemplated as this helps to delay rotting of harvested tubers.

There are various home techniques which are practised to produce storable material including sun drying or making flour, tapioca or other starchy preparations. Stock feed preparations are also prepared from the sweet cassavas.

Pests and pathogens

The pests and pathogens of cassava are listed in Table 13.5.

Further Reading

Araceae

Food and Agriculture Organization (FAO) (2010) *Quality Declared Seed for Vegetatively Propagated Crops.* FAO Plant Production and Protection Paper. FAO, Rome, Italy.

Lebot, V. (2009) *Tropical Root and Tuber Crops Cassava, Sweet Potato, Yams and Aroids.* CAB International, Wallingford, Oxon, UK, pp. 277–360.

Revill, P.A., Jackson, G.V.H., Hafner, G.J., Yang, I., Maino, M.K., Dowling, M.L., Devitt, L.C., Dale, J.L. and Harding, R.M. (2005) Incidence of distribution of viruses of taro (*Colocasia esculenta*) in Pacific Island countries. *Australasian Plant Pathology* 34, 327–331.

Convolvulaceae

Ames, T., Smit, N.E.J.M., Braun, A.R., O'Sullivan, J.N. and Skogland, L.G. (1996) *Sweet Potato: Major Pests, Diseases and Nutritional Disorders*. International Potato Center (CIP), Lima, Peru.

Hall, A.J. and Devereau, A. (2000) Low cost storage of fresh sweet potatoes in Uganda: lessons from participatory and on station approaches to technology choice and adaptive testing. *Outlook on Agriculture* 29(4), 275–282.

Lebot, V. (2009) *Tropical Root and Tuber Crops Cassava, Sweet Potato, Yams and Aroids*. CAB International, Wallingford, Oxon, UK, pp. 90–179.

McDonald, J.A. and Austin, D.F. (1990) Changes and additions in *Ipomoea* section Batatus (Convolvulaceae). *Brittonia* 42, 116–120.

O'Sullivan, J., Asher, C.J. and Blamey, F.P.C. (1997) *Nutrient Disorders of Sweet Potato*. ACIAR Monograph No. 48, Canberra, Australia.

Palada, M.C. and Chang, L.C. (2003) *Suggested Cultural Practices for Kang Kong*. Asian Vegetable Research and Development Center (AVRDC) Publication 03-554. AVRDC, Taiwan.

Scott, G.J. (2005) *Sweet Potatoes as Animal Feed in Developing Countries: Present Patterns and Future Prospects*. International Potato Center (CIP), Lima, Peru.

Stathers, T., Namanda, S., Mwanga, R.O.M., Khisi, G. and Kapinga, R. (2005) *Manual for Sweet Potato Integrated Production and Pest Management for Farmer Field Schools in Sub-Saharan Africa*. International Potato Center (CIP), Kampala, Uganda.

Dioscoraceae

Coursey, D.G. (1967) *Yams*. Tropical Agriculture Series. Longman, Harlow, UK.

Lebot, V. (2009) *Tropical Root and Tuber Crops Cassava, Sweet Potato, Yams and Aroids*. CAB International, Wallingford, Oxon, UK, pp. 181–275.

Proctor, F.J., Goodliffe, J.P. and Coursey, D.G. (1981) Post-harvest losses of vegetables and their control in the tropics. In: Spedding, C.R.W. (ed.) *Vegetable Productivity*. Macmillan, London, pp. 139–172.

Revill, P.A., Jackson, G.V.H., Hafner, G.J., Yang, I., Maino, M.K., Dowling, M.L., Devitt, L.C., Dale, J.L. and Harding, R.M. (2005) Incidence of distribution of viruses of taro (*Colocasia esculenta*) in Pacific Island countries. *Australasian Plant Pathology* 34, 327–331.

Cassava

Food and Agriculture Organization (FAO) (2010) *Quality Declared Seed for Vegetatively Propagated Crops*. FAO Plant Production and Protection Paper. FAO, Rome, Italy.

Hillocks, R.J., Thresh, J.M. and Belloni, A. (2002) *Cassava: Biology, Production and Utilization*. CAB International, Wallingford, Oxon, UK.

Lebot, V. (2009) *Tropical Root and Tuber Crops Cassava, Sweet Potato, Yams and Aroids*. CAB International, Wallingford, Oxon, UK, pp. 2–179.

Rawel, K.M. and Kroll, J. (2003) The importance of cassava (*Manihot esculenta* Crantz) as the main staple food in tropical countries. *Deutsche Lebensmittel-Rundschau* 99(3), 102–111.

14 Andean Tubers and Roots and Crops of the *Lamiaceae* and *Apiaceae*

This chapter covers the eight crop species known collectively as Andean tubers and roots. Also included at the end of this chapter are other vegetables in the *Lamiaceae* (formerly *Labiatae*) and *Apiaceae* (formerly *Umbelliferae*); these crops are not regarded as Andean tubers.

Andean Tubers and Roots

The Andean tuber and root group is generally considered to include the following eight species:

- *Oxalis tuberosa* Molina in the family *Oxalidaceae* – oca (common name);
- *Ullucus tuberosus* Caldas in the family *Basellaceae* – ulluco, ulluca, melloco;
- *Tropaeolum tuberosum* Ruiz and Pavon in the family *Tropaeolaceae* – mashwa, mashua;
- *Mirabilis expansa* Ruiz and Pavon in the family *Nyctaginaceae* – mauka, chago;
- *Lepidium meyenii* Walp. in the family *Cruciferae* – mace, peppergrass, maca;
- *Solanum* x *juzepczukii* and *Solanum* x *curtibolum* in the family *Solanaceae* – bitter potatoes, ruku, choquepico, luki;
- *Arracacia xanthorrhiza* Bancroft in the family *Apiaceae* (formerly *Umbelliferae*) – arracacha, white carrot, Peruvian parsnip; and
- *Canna edulis* Ker. in the family *Cannaceae* – achira, edible canna, purple arrowroot, Queensland arrowroot.

The Andean tuber crops comprise a very interesting and long-recognized group of eight vegetables selected and grown over many centuries by subsistence farmers in the Andes during and since Inca times. The continual selection and

 © R.A.T. George 2011. *Tropical Vegetable Production* (R.A.T. George)

maintenance of this group of root vegetable species has enabled farmers to survive in areas of low rainfall, high altitude with high UV light, rugged topography, poor and shallow soils. The food values, biology, distribution, history and biochemistry of this crop group are described by Flores *et al.* (2003). Their genetic diversity, ecology and prospects for improvement have been outlined by Arbizu and Tapia (1994).

Oca, ulluco and mashwa are the three most widely grown species within the diverse group providing an important part of the diet in the Andean regions of South America, from Venezuela to Argentina, and parts of Mexico. Oca has also been considered as a potential food crop for temperate areas of the world and is already cultivated in New Zealand. These three crops are sometimes cultivated in mixed companion plantings in South America, especially in Peru; subsistence farmers often include bitter potatoes in this mixed planting.

Oxalis tuberosa, Oxalidaceae (oca)

This is a perennial, generally cultivated annually from new plant propagules. The stock is maintained by on-farm selection and vegetative propagation of its conical tubers which have 'eyes' on most surface areas. All parts of the plant are eaten; surplus roots can be dried and stored for long periods as 'kaya'. The crop succeeds in relatively dry conditions in a soil from pH 5.5 to neutral.

The new season's planting material is often produced by division of tubers selected by farmers from their on-farm material. However, attempts are now being made to improve the quality of planting materials (for oca, ulluco and mashwa) by providing protocols for improving stocks, especially in relation to the transmission of pests and pathogens via the planting materials (FAO, 2010).

The major viruses of oca are listed in Table 14.1.

Ullucus tuberosus, Basellaceae (ulluco, ulluca, melloco)

This is a perennial herbaceous species whose aerial shoots grow up to 0.5m but with a tendency to scramble. The tubers are produced on underground stems, varying in outer skin colour according to their original clone; internal flesh colour is yellow in all selections. The species has a high tolerance of diurnal temperature variation and many selections withstand frost.

Tubers or propagules from selected tubers are planted in shallow drills 50–80cm apart and earthed up. The crop is earthed up again when aerial shoots emerge, further earthing up assists in weed control and tuber formation. Tubers can be harvested 4–5months from planting.

The mature tubers are used as a cooked vegetable and the leaves are also boiled for culinary use. The tubers do not store well, but any surplus can be dried and the resulting 'lingle' stored for long periods and used as flour.

Stone (1982) has described techniques for the elimination of four viruses from ulluco using meristem culture and chemotherapy. The pests and pathogens of ulluco are listed in Table 14.2.

Table 14.1. Major viruses of oca.

Virus	Mode of transmission
Potato black ring spot virus (PBRSV)	Nematodes
Arracacha virus B (AVB-O)	Nematodes
Potato virus T (PVT)	Usually by mechanical contact
Papaya mosaic virus (PapMV-U)	Usually by mechanical contact

Table 14.2. Major pests and pathogens of ulluco.

Pest or pathogen	Common names
Premnotrypes solani	Andean weevil
Amathynetoides nitidiventris	Ulluco weevil
Meloidogyne spp.	Root knot eelworm
Myzus persicae	Aphid
Potato leaf roll virus (PLRV)[a]	
Potato virus T (PVT)[b]	
Papaya mosaic virus (PapMV-U)[b]	
Ulluco mosaic virus (UMV)[c]	
Ulluco mild mosaic virus (UMMV)[b]	

[a] The vectors of PLRV are *Aphis* spp. and *M. persicae.*
[b] PVT, PapMV-U and UMMV are usually transmitted by mechanical contact.
[c] The vector of UMV is *M. persicae.*

Tropaeolum tuberosum, Tropaeolaceae (mashwa, mashua)

This is a perennial although a new crop is usually produced annually from true seed or vegetatively propagated by tuber cuttings. The main viruses of mashwa and their vectors are listed in Table 14.3.

Mirabilis expansa, Nyctaginaceae (mauka, chago)

This is an herbaceous perennial of up to 1 m in height with edible foliage and a storage root. It is generally grown as an annual, although farmers may let it remain in the field as a means of storage or an extension of the cropping season. It produces a crop at temperatures ranging from 4 to 30°C. The edible roots vary in their external colour and change from yellow to white according to age; their length can be up to 40 cm.

Lepidium meyenii, Crucifereae (mace, peppergrass, maca)

This is an herbaceous biennial with an edible, spherical hypocotyl, resembling a radish. It can produce a useful annual crop, even at altitudes of up to 4500 m. The harvested hypocotyls are dried and can be stored for several years.

    The crop is usually produced from true seed sown prior to the predicted rains. Farmers frequently grow the crop in rows alternating with bitter potatoes because

Table 14.3.   Major viruses of mashwa and their vectors.

Pest or pathogen	Mode of transmission
Sowbane mosaic virus (SOMV)	Leaf miners, beetles and *Aphis* spp.
Tropaeolum mosaic virus (TrMV)	*Aphis* spp.
Potato virus T (PVT)	Usually by mechanical contact
Papaya mosaic virus (PapMV-U)	Usually by mechanical contact

it is believed that this pattern of multi-cropping reduces insect damage and subsequent yield reduction of the bitter potatoes.

Solanum x *juzepczukii* and *S.* x *curtibolum, Solanaceae* (bitter potatoes, ruku, choquepico, luki)

Origins and types

The 1920's expedition organized by Vavilov described these two species; their specific epithets are complimentary to the two students who accompanied him (Arbizu and Tapia, 1994). The two bitter potatoes have been domesticated for approximately 8000 years and are important subsistence farmer crops in the high altitudes of Peru and Bolivia.

Crop production

The growing cycle is from 5 to 8 months, depending on the clone. The plants produce higher yields in soils with an organic content of up to 5%. Fallow periods of up to 6 years coinciding with applications of animal manure from grazing will assist soil improvement. A rotation which includes a fodder crop and *Chenopodium pallidcaule*, (known as 'one of the lost crops of the Incas' common name kaniwa) also increases soil organic matter content (Arbizu and Tapia, 1994). Kaniwa flour is used as a bread preparation. Some cropping patterns include mixed plantings with oca and ulluco.

The crop is grown under rain-fed conditions; staggered plantings are made as an insurance to allow for variations in the start of the rainy season. The crop is earthed up several times following emergence of aerial shoots. Tuber formation commences at the start of anthesis which is approximately 2 months from emergence. The main pests and pathogens are *Naccobbus aberrans* (a nematode), *Premnotrypes* spp. (Andean weevil) and *Synchytrium endobioticum* (the cause of wart disease).

The bitterness of these two *Solanum* species is due to glycoalkaloids which must be removed before eating; this is achieved by exposure to low temperatures and air drying in sunlight (see also Arbizu and Tapia, 1994).

Arracacia xanthorrhiza, Apiaceae (arracacha, white carrot, Peruvian parsnip)

This is an important root crop in the Andean Highlands, and has also been regarded as a species of value for other tropical highlands, including Africa

(National Academy of Sciences, 1978). Arracacha is a semi-prostrate perennial although it is normally cultivated as an annual and usually reproduced by propagation from off-shoots derived from the main root crown.

The leaves and younger stems are edible, hence it is often referred to as 'Creole celery'. Other parts of the plant are also valued including the main tuber and secondary tubers. After harvesting these different morphological parts the main root crown and coarser older leaves are used as stock feed.

The following season's crop is produced by rooting the short shoots arising from attached off-sets which emerge from the main root crown. The plant produces seed in some environments but as the pathogen *Septoria apii-graveolentis* (cause of leaf spot, otherwise known as late blight) is seed transmitted, vegetative propagation is strongly recommended. Propagation material should be selected from plants which appear to be disease and pest free. The unrooted cuttings are usually planted direct into their final positions, although some farmers prefer to make cross-cuts in the base of the propagules and dry the prepared material for 2 or 3 days before planting. The cuttings are planted at the start of the rains and produce a crop in approximately 1 year. The spacing and earthing-up sequences are as described above for ulluco.

Major pests and pathogens of arracacha are:

- *Epitrix* spp. – flea beetle;
- *Cercospora* spp. – cause of leaf spot;
- *Erwinia* spp. – cause of bacterial rot; and
- *Septoria apii-graveolentis* – cause of leaf spot (late blight).

Canna edulis Ker. *Cannaceae* (achira, edible canna, purple arrowroot, Queensland arrowroot)

Achira has been an important food crop in the Andes for centuries. The large, branched rhizomes (up to 60 cm long) are baked and eaten, the leaves and tubers are also used for livestock feed. The mature plant grows to approximately 1 m high and succeeds at high altitudes. The crop is also produced by small farmers in parts of Asia and the Pacific Islands where it is used for commercial noodle production. The species is cultivated commercially in Australia for starch production where it is commonly known as Queensland arrowroot.

The crop is propagated from terminal tuber cuttings which are planted and cultivated as described for ulluco. The crop matures in approximately 8 months at low altitudes but can take up to a year at the higher altitudes.

Lamiaceae (Formerly *Labiatae*)

Coleus parviflorus Benth. (syn. *C. tuberosus* Benth.) (Sudan potato)

The Sudan potato is mainly cultivated in tropical Africa (especially Nigeria, the Sudan and Upper Volta), India, Indonesia and Malaysia. The species is an herbaceous perennial reproduced annually from selected tubers. The slightly aromatic tubers are produced below ground close to the stem. The tubers are similar in shape to those of *S. tuberosum* but generally smaller. There are local selections based on external tuber colour of black, red or white, but all have white internal flesh.

The 'seed' tubers produce shoots during their storage period and it is important that these are not accidentally knocked off or damaged during transportation and planting. The sprouting tubers are usually planted in rows 50 cm apart, with 30 cm between planting stations. The rows are initially earthed up, which is repeated in the early establishment of the crop to provide a soil environment for tuber production, especially developing tillers, also for weed control. On sites prone to waterlogging 'seed' tubers are planted on mounds or ridges.

The Sudan potato responds to organic manures which are incorporated within the ridges at planting; a side dressing of a balanced N:P:K fertilizer applied at a rate of $40 \, g/m^2$ after emergence would also be expected to increase yield provided that there are no other limiting factors.

The crop can be planted at any time during the year and matures in 3–7 months, depending on location; irrigation is necessary if the growing period includes the dry season. Tuber maturity follows anthesis after which the foliage senesces.

The mature tubers do not store well if left in the ground but are lifted and dry stored in containers of dry sand, soil or dry debris such as husks from seed threshing. In some areas it is essentially a crop produced to store for food security during the dry and non-productive seasons. However, as the tubers lose their culinary quality after approximately 8 weeks from lifting (coinciding with the start of sprouting) the tubers are parboiled followed by air drying for longer term storage.

Crop and yield improvement can be achieved by selection of larger, insect- and pathogen-free tubers for the following season's planting, also stem cuttings and tissue culture; the protocols for their handling, production and quality standards are outlined by FAO (2010). Tubers selected for planting are stored in dry materials as outlined above. The 'seed' tubers should be kept dry and out of reach of storage pests.

Solenostemon rotundifolius (Poir.), J.K.Morton (hausa potato, frafra potato, ketang, country potato)

This species is a member of the *Lamiaceae* (formerly *Labiatae*) but is not regarded or included as an Andean tuber crop.

Origins and types

Hausa potato is an annual herbaceous species regenerated from tubers. It grows to a maximum height of approximately 30 cm. It is believed to have originated in East Africa and became an important tuber vegetable in much of Africa. Although it has now been largely replaced by the introductions of cassava and sweet potato from South America it remains a very important crop for subsistence farmers in the drier or arid areas of Africa, including Northern Ghana, Mali and the savannah areas of the Sudan where it remains a favoured crop. The species is also cultivated in parts of South-east Asia.

The edible tubers are morphologically similar to those of *S. tuberosum* but smaller. Clones or local types exist which have a range of skin pigmentations, including white, red, brown and black; the internal tuber colour can be white, red, brown or black.

Cultural requirements

Hausa potato can tolerate areas of high rainfall but is intolerant of waterlogging; cool nights tend to induce tuber development. Well-drained sandy loams are ideal for this crop although being a species that will produce a crop under adverse conditions it will survive in very poor soils, hence its value for subsistence farmers.

Propagation

Tubers which have commenced sprouting, or shoots severed from tubers, are used for planting material. Selected tubers for propagule production are shallow planted up to 5 cm deep and at a spacing of approximately 8×8 cm in prepared beds which have had bulky organic manures or compost incorporated. The emerging shoots are removed from the mother tubers when approximately 15 cm high; shoots can take up to 8 weeks before suitable cutting material has started to be produced. In some areas the beds are maintained for further cutting production providing continuity of supply in suitable environments. A proposed protocol for the production of improved quality hausa potato planting materials is described by FAO (2010).

Planting and crop maintenance

The propagules are planted in prepared ridges 80 cm apart with 15–20 cm within the rows. There are several ways of placing the cuttings, from vertical to horizontal, but generally the method used is to place the cuttings vertically with approximately 5 cm of the basal end inserted in the soil. The soil surface is kept weed free for the initial 4–5 months and the ridges are earthed up at the final hoeing. The crop benefits from a top dressing of an N:P:K 2:1:1 inorganic fertilizer applied after establishment at a rate of 30 g/m^2.

Harvesting

Groups of tubers are produced close to the bases of shoots and are ready for harvest in approximately 5–8 months from planting, depending on location and local clone. The plants start to senesce by this stage and tubers which are left in the soil will rot.

The harvested tubers are often mixed with dry sand or seed dross and stored in containers such as large pots which are often suspended in rooms or sheds to reduce the risk of pest damage. Otherwise the tubers are spread on a bed of dry leaves in the shade and covered.

Pests and pathogens
Very few pests or pathogens are reported in the literature, although *Meloidogyne* spp. (root knot eelworm) is a widespread problem. Other pest-associated problems arise from feral pigs and termites.

Apiaceae

The following crop is a member of the *Apiaceae* (formerly *Umbelliferae*) but is not considered to be an Andean tuber.

Daucus carota L. subsp. *sativus* (Hoffm.) Thell. (carrot)

Origins and types
The modern cultivated carrot has been derived from the wild carrot *Daucus carota* L. found in Europe, Asia and Africa. The crop is cultivated in parts of Africa, areas of tropical Asia, South America and the Caribbean where it succeeds at altitudes greater than 500 m. The cultivated carrot is a biennial normally producing an edible swollen tap root in the first year and flowering in its second year; as a root vegetable it is cultivated as an annual to be harvested in its first year. A very small percentage of plants may flower ('bolt') in their first year and although these can be fed to livestock they should not be used for 'on-farm' seed production. The biennial roots and foliage ('tops') can also be fed to livestock if necessary.

There are two main sources of material from which early cultivars were developed: (i) the anthocyanin carrots from Asia (especially centred in Afghanistan where purple-rooted carrots can still be found in cultivation); and (ii) the carotene carrots, with their swollen tap roots which were further developed in Europe. Banga (1963) described the descent and development of the main types of the 'Western carotene carrot'. This is the source of material from which cultivars grown in the tropics and elsewhere have been developed and further improved by plant breeders to have a high content of carotene in the root and its internal core (in addition to other desirable characters). There are open-pollinated and F_1 cultivars. The carrot seed of commerce has normally had its appendages removed by a de-bearding machine during seed processing; the purpose of this is to ensure that the seeds do not form clusters which interfere with the free flow of seed during further processing and sowing.

Carrot cultivars are usually classified according to maturity period, root shape (length and width), shape of root tip and foliage. Further divisions can be made according to their season of use and root colour. Banga (1963) has described the types of orange-rooted carrots and their specific characters. The six generally recognized root types of carrot are illustrated in Fig. 14.1. The short-rooted cultivars succeed in the lowland tropics; the longer Nantes types are more suitable in the higher tropical elevations.

Fig. 14.1. Morphological types of carrot roots (left to right): Short horn, Nantes, Amsterdam forcing, Chantenay, Autumn King and St Valery.

Required environment

Successful root crops for culinary and livestock can be produced provided the soil temperature remains below approximately 25°C. Lower root carotene levels tend to occur at higher temperatures.

Soil

Carrots are most successful in a deep loam; root production in well-drained sandy soils can also be very successful provided there is sufficient available water throughout the crop's life. Stony soils should be avoided as much as possible. Carrots do not tolerate acid conditions: the optimum pH is 6.5, soils with a lower pH should have an application of a liming material during preparations.

Recently manured land should not be used for carrot production as it tends to cause misshapen roots (often referred to as 'fanging') However, sites which received bulky organic manures for a previous crop, such as onions are suitable.

Site preparation

The crop should be produced on raised beds if waterlogging is likely to occur. The crop can be produced in beds or as a row crop. Weed control is an important issue and when possible the stale seedbed technique should be applied, although the seed normally germinates in approximately 10 days, assuming seed of acceptable vigour and germination has been used.

Table 14.4. The main pests and pathogens of *D. carota* (carrot).

Pest or pathogen	Common names
Ditylenchus dipsaci	Eelworm
Meloidogyne spp.	Root knot eelworm
Alternaria dauci	Carrot or black leaf blight
Alternaria radicina	Black root rot, seedling blight
Cercospora carotae	Cercospora blight of carrot, leaf spot
Corticium solani	Leaf blight
Erysiphe heraclei	Powdery mildew
Phoma rostrupii	Phoma root rot
Sclerotium rolfsii	Wilt, collar rot
Xanthomonas campestris pv. *carotae*	Bacterial blight, root scab
Aster yellows virus	
Carrot motley dwarf virus (a complex of three viruses, including carrot red leaf virus)	

Care must be taken in the preparation of carrot seedbeds to produce a satisfactory tilth of final particle size ranging from rice grain to pea in sizes. If the final tilth is too fine it will pan and form a crust and if too coarse there will be seedling emergence problems; a poorly prepared seedbed can result in malformed roots.

Sowing

The seed is sown in rows 25 cm apart, or broadcast, at a depth of 1 cm. A light mulch of chopped straw or similar material should be applied after sowing to prevent soil capping and also assist the reduction of subsequent soil temperatures.

The emerging seedlings should be thinned as soon as they can be handled to approximately 3–5 cm apart, depending on the cultivar. If left at a too high plant density many roots will be distorted or misshapen and intraspecific competition will reduce the useful root yield. Using coated or pelleted seed assists in obtaining an even seedling stand and should minimize the need for post-emergence thinning. Erecting a lightweight shade structure over the crop can improve quality in high temperature environments.

Harvesting

Carrots take approximately 3 months to mature in the tropics; the earlier, short-rooted cultivars take less time to reach maturity than the longer rooted cultivars. Subsistence farmers should use the crop from the ground as required, unless there is a niche market locally to justify a bulk harvest. The root crop does not store well except at lower temperatures and this is not normally economic in the tropics.

The main pests and pathogens of carrot are listed in Table 14.4.

The minor crops in *Apiaceae*

These species are generally of local importance to subsistence farmers and are generally cultivated as pot herbs, salads, flavourings, garnishes or condiments.

They are usually grown from seed in small amounts, unless there is a local niche market. This group includes:

- *Anethum graveolens* L. – dill;
- *Anethum sowa* Kurz. – Indian dill;
- *Anthriscus sylvestris* Hoffm. – chervil;
- *Carum carvi* L. – caraway;
- *Coriandrum sativum* L. – coriander;
- *Cuminum cyminum* L. – cumin;
- *Foeniculum vulgare* Mill. – fennel;
- *Foeniculum vulgare* Mill. var. *dulce* (Mill.) Thell. – Florence fennel; and
- *Pimpinella anisum* L. – anise.

Some species in this family are also important sources of essential oils, especially the dills, celery, caraway and cumin.

Further Reading

Arbizu, C. and Tapia, M. (1994) Andean tubers. In: Hernado Bermejo, J.E. and Leon, J. (eds) *Neglected Crops: 1492 from a Different Perspective.* Plant Production and Protection Series No. 26. Food and Agriculture Organization (FAO), Rome, Italy, pp. 149–163.

Flores, H.E., Walker, T.S., Guimaraes, R.L., Bais, H.P. and Vivanco, J.M. (2003) Andean root and tuber crops: underground rainbows. *HortScience* 38(2), 161–167.

Food and Agriculture Organization (FAO) (2010) *Quality Declared Seed for Vegetatively Propagated Crops.* FAO Plant Production and Protection Paper. FAO, Rome, Italy.

Gade, D.W. (1966) Achira, the edible canna – its cultivation and use in the Peruvian Andes. *Economic Botany* 20, 407–415.

Lebot, V. (2009) *Tropical Root and Tuber Crops Cassava, Sweet Potato, Yams and Aroids.* CAB International, Wallingford, Oxon, UK.

Leon, J. (1964) La maca (*Lepidium meyenii*) a little known food plant of Peru. *Economic Botany* 18, 122–127.

National Academy of Sciences (1978) II Roots and tubers. In: *Unexploited Tropical Plants with Promising Economic Value.* National Academy of Sciences, Washington, DC, pp. 29–43.

Stone, O.M. (1982) The elimination of four viruses from *Ullucas tuberosus* by meristem culture and chemotherapy. *Annals of Applied Biology* 101, 79–83.

15 *Gramineae* and *Cyperaceae*

Gramineae

The *Gramineae* family contains many extremely important food crops for both human and livestock nutrition and contains economically important agricultural grain and cereal crop species. These crops fall into three categories: (i) the cereal grain crops; (ii) the tropical grasses; and (iii) the grain and vegetable crops exemplified by *Zea mays* L.

Cereal grain crops

These include the following species:

- *Triticum* species – wheats;
- *Oryza sativa* L. and its subspecies – rice, paddy;
- *Oryza glaberrima* Steudel – African rice;
- *Hordeum vulgare* – barley;
- *Avena sativa* L. – oat;
- *Sorghum bicolor* (L.) Moench. – sorghum;
- *Pennisetum glaucum* (L.) R. Br. emend Stuntz – pearl millet, bulrush millet; and
- *Eleusine coracana* (L.) Gaertn. – finger millet.

Tropical grasses

These may be used for: (i) 'grassing down' in crop rotations; (ii) break crops following serious soilborne pest or pathogen infections when it is prudent to cease the production of susceptible vegetable groups; or (iii) for grazing and/or cut fodder and hay.

They include:

- *Chloris gayana* Kunth. – Rhodes grass – successful on a wide range of soil types, although not very drought resistant.
- *Setaria sphacelata* (Schum.) Staph *et* C.E. Hubb. – setaria, South African pigeon grass, golden timothy – a tufted perennial, sometimes with rhizomes, grows up to 1 m.
- *Panicum coloratum* L. – small buffalo grass, kleingrass – this species sheds its seed very readily and may possibly produce problems of 'volunteers' in following arable crops.
- *Sorghum* x *almun* Parodi – Colombus grass.
- *Sorghum sudanensc* (Piper) Slaph. – Sudan grass.
- *Sorghum* x *sudanense* – Sudan grass – this species is an improved form of the original Sudan grass which originated from the Sudan and South Africa. The improvement has been mainly made by crossing *S. sudanense* with *Sorghum bicolor.*
- *Bracharia decumbens* Staph. – signal grass, Surinam grass – native to tropical Africa, it requires humid conditions.
- *Cenchrus cilaris* L. – buffelgrass – suitable for dry tropical areas.
- *Hyparrhenia rufa* (Nees) Staph. – jaragua grass, thatching grass, puntero, veyale – this species can grow from 60 to 240 cm in height. It is widely grown in Central and South America and also tropical Africa. It is able to tolerate seasonable flooding but not waterlogging; in the Ilanos of Bolivia and Colombia it tolerates and survives up to 6 months of dry season. This grass is an important thatching material in Africa where it is also used as a straw for livestock.
- *Melinis minutiflora* P. Beauve. – molasses grass – survives well on poor soils although requires satisfactory drainage.
- *Paspalum plicatulum* Michaux. – plicatulum – a tall perennial grass, successful on most soils in the medium to high rainfall tropics.
- *Urochloa mosambicensis* (Hackel) Dandy. – sabi grass – this grass is cultivated in the semi-humid tropics. It requires good drainage but it can withstand drought and heavy grazing.
- *Panicum maximum* Jacq. – Guinea grass – some cultivars can grow to a height of approximately 3 m.
- *Paspalum dilatatum* Poiret. – dallis grass, paspalum – this is a perennial and is a native of tropical America.
- *Paspalum notatum* Flugge. – bahia grass – this species originates in tropical America, is relatively deep rooted and forms a firm sod (turf) relatively quickly.
- *Pennisetum clandestinum* Hochst. Ex Chiov. – kikuyu grass – this is a rhizomatous grass that can withstand heavy grazing.
- *Saccharum officinarum* L. – this is generally regarded as the predominant species cultivated for the production of sugarcane. Purseglove (1972) provides an interesting account of the range of *Saccharum* species used for their sweetness and sugar content, their origins, distribution and agronomy.

Zea mays (sweetcorn, corn on the cob, vegetable maize, mealies, corn)

According to Purseglove (1972) the cultivars of *Z. mays* can be classified into seven groups:

1. Pod corn – Z. *mays tunica* Sturt.
2. Popcorn – Z. *mays everata* Sturt. (syn. *praecox*).
3. Flint maize – Z. *mays indurata* Sturt.
4. Soft or flour maize – Z. *mays amylacea* Sturt. (syn. *erythrolepis*).
5. Sweetcorn – Z. *mays saccharata* Sturt. (syn. *sugosa* Bonof.).
6. Waxy maize – Z. *mays ceritina* Kulesh.

There are also ornamental forms, for example: (i) var. *japonica* Koern. which has striped leaves; and (ii) var. *gracillima* Koern., a dwarf form. The different cultivars of these groups cross-pollinate which can sometimes make it difficult to classify them exactly.

The two crops considered as vegetable crops outlined here are:

- sweetcorn; and
- baby corn (also known as mini-corn).

There are two types of sweetcorn cultivars which have a higher proportion of sugars to starches in the seeds' endosperm than other sweetcorn cultivars. These two types are: (i) super sweet; and (ii) sweet. The super sweet has a genetically controlled factor for extra sugar in the seeds resulting in a longer period for the change to starch after harvesting; they are therefore sweeter tasting at harvest and remain sweeter postharvest than the sweet cultivars, although the sweet cultivars are sweeter than the cultivars grown for grain. The several endosperm mutations of sweetcorn have been described by Wolfe *et al.* (1997). They include sugary1 (su1), shrunken2 (sh2), sugary enhancer1 (se1), brittle (bt2), amylose extender (ae), dull (du) and waxy (wx). There is a large range of both open pollinated and F_1 hybrid cultivars.

Site, soil and nutrition

None of the Z. *mays* types grow well in shady conditions. Sweetcorn tolerates a range of soil types, but it is important that the soil should have satisfactory water retention ability. Where possible the addition of bulky organic manures or materials will assist soil water retention. This crop tolerates a soil pH of 5.5–6.8; appropriate quantities of lime or a liming material should be applied if the pH is below this range but it should not be added with the organic manures. Some growers apply a higher proportion of nitrogen in the base dressing while others may give a nitrogenous top dressing during the young plant stage if suitable fertilizers are available. A balanced compound fertilizer incorporated at a rate of $40\,g/m^2$ during final preparation of the site will be beneficial. If a nitrogenous fertilizer is indicated during the young plant stage it should be applied as a top dressing at a rate of $20\,g/m^2$.

Sowing and crop establishment

Zea mays is wind pollinated so if flowering occurs during periods of heavy rain, pollination and seed setting in the cobs is likely to be sparse, leading to low quality cobs and low yield; therefore time of sowing should be scheduled to correspond with satisfactory predicted local conditions. Depending on the cultivar and local conditions the crop matures approximately 3–4 months from sowing.

The crop can be grown on either raised beds, ridges or on the flat. The developing plant produces adventitious 'prop' roots from its basal stem nodes; these prop roots provide further support for the plant. Bearing in mind the potential of the prop roots, growing on the flat may be considered to be the best system unless the required irrigation method dictates otherwise.

Seed is sown direct at a depth of approximately 1.5–2.5 cm, the greater depth is used for larger seeded cultivars (many of the sweet cultivars tend to have smaller seeds). The rows are 60 cm apart. After seedling emergence the plants are thinned to 30–36 cm apart within the rows, depending on the potential vigour of the cultivar.

Early weed control is important with *Zea* crops (except for a high density crop grown exclusively for cutting as fodder).

The plants are usually earthed up when they are approximately 20 cm high. This operation is normally done while hoeing for weed control. Further earthing up can be done during subsequent hoeing, but care must be taken to avoid root damage by deep and vigorous hoeing.

Irrigation

Some sweetcorn crops are produced in areas with relatively low summer rainfall so it is important that sufficient irrigation is available. MacKay and Eaves (1962) found that sweetcorn was very responsive to supplementary irrigation especially from the pollen-shedding stage to cob maturity; also that removing the moisture stress by increasing irrigation increased the crop's response to nitrogen, phosphorus and potassium. The evidence for sweetcorn response to water availability is generally in accord with research findings of maize.

Wolfe *et al.* (1988) reported that water stress during anthesis adversely affected the synchrony of pollen release and the receptiveness of the 'silks' (female flower parts).

Harvesting

Sweetcorn cobs are normally harvested while the developing grain is still 'milky' (i.e. before the grain starts to harden) which is observed when the maturing grain is slightly pressed with a finger or thumb nail; a useful preliminary indication of this stage is when the silks at the ends of cobs have started to wither and become partially dry. The cob is not ready for harvest while the juices from the squashed grain are still clear. The level of maturity is usually confirmed by looking under the husks of a small number of sample cobs while they are still attached to the mother plants. Developing seeds of sweetcorn cultivars are more translucent at this stage than maize (grain) cultivars.

Cobs are usually harvested as required, but unlike maize, sweetcorn should be harvested before the cobs' husks start to turn from green to brown and dry. When not for immediate use, the harvested cobs should be kept cool and shaded. Depending on prevailing temperatures cobs can be stored for up to 1 week, but as the sugar levels reduce after harvesting they should ideally be harvested as required.

Production of baby corn

Baby corn is a very traditional Chinese crop which has more recently become a popular crop in other parts of Asia including Indonesia and the Philippines. It has also become an important large-scale commercial crop for export to Europe and other major world markets in temperate areas. The crop can be produced from the sweet or super sweet cultivars, although as it is the immature and unfertilized cobs which are harvested many growers use open-pollinated maize cultivars, the seed of which is cheaper per unit weight.

The site, soil and preparation are as described above for sweetcorn. The crop is produced on the flat, on ridges or in furrows depending on the soil type, drainage, predicted rainfall and irrigation method.

Seed is sown on a weed-free prepared plot, at a depth of approximately 3–5 cm. Sowing depth is increased for the larger seeded maize cultivars, in rows approximately 1 m apart, with approximately 25 cm between sowing stations within the rows. Many farmers sow two to three seeds per station but do not normally thin out the seedlings.

Weed control is especially important during the early stages of crop establishment. This is usually achieved by hoeing, although when the crop is grown on ridges care must be taken not to undermine the plants and deter the production of 'prop' roots.

The plants are detasseled as soon as the 'tassels' (male inflorescences) emerge from the tops of the plants, this can be done with a sharp knife, shears or secateurs. The object of this task is to ensure that the immature cobs are not pollinated (i.e. the harvested baby corn is maternal material only).

The baby corn is harvested approximately 3 days after silking ('silking' is the appearance or emergence of the styles which collectively form the silks of individual female inflorescences); in many cultivars this is approximately 7–8 weeks from sowing. The overall harvesting period is approximately 2 weeks, depending on the uniformity of the crop and the homozygosity of the seed lot sown. Early cultivars can be considered an advantage over those which take longer to reach the silking stage. Another character to take into consideration is plant height, depending on the secondary use of the plant remains after harvest, such as composting, digging in or fodder.

The baby corn is harvested early in the mother plant's life and therefore the crop is not exposed to the full range of pests and pathogens as the main maize crop in its cropping cycle.

The young cobs are carefully broken off the mother plant and the protective leaves and immature husk removed when required for culinary use. From a subsistence farmer's point of view the baby corn's cropping cycle has fewer pests and pathogens than a maize or sweetcorn crop grown to maturity, thus by producing baby corn the overall risk and costs relating to crop protection are reduced. An additional consideration is that because the crop is consumed in a very young and immature stage it can be relatively free of pesticide residues.

Pests and pathogens

The main seedborne pathogens of *Z. mays* and its subspecies are listed in Table 15.1.

Table 15.1. The main pests and pathogens of *Z. mays*.

Pest or pathogen	Common names
Heliothis armigera	Corn ear-worm
Laphygma frugiperda	Corn leaf-worm
Acremonium strictum	Kernel rot
Cochliobolus carbonum	Charred ear mould, southern leaf spot
Cochliobolus heterostrophus	Southern leaf spot, blight
Colletotrichum graminicola	Anthracnose
Diplodia frumenti	Dry ear rot, stalk rot, seedling blight
Fusarium spp.	Fusarium
Gibberella fujikuroi	Gibberella ear rot, kernel rot, stalk rot
G. fujikuroi var. *subglutinans*	Seedling blight
Gibberella zeae	Seedling blight, cob rot
Marasmius graminum	Seedling and foot rot
Penicillium spp.	Seed rot, blue-eye
Physoderma maydis	Brown leaf spot
Puccinia polysora	Rust
Sclerophthora macrospora	Crazy top
Stenocarpella macrospora and *Stenocarpella maydis*	Dry or white ear rot, stalk rot, seedling blight, root rot
Ustilaginoidea virens	False smut, green smut
Ustilago zeae	Smut, blister or loose-smut
Erwinia stewartii	Bacterial leaf blight, bacterial wilt, Stewart's disease, white bacteriosis
Maize leaf spot virus	
Maize mosaic virus (MMV)	
Sugarcane mosaic virus (SMV)	

Cyperaceae

Cyperus esculentus Linn. (earth nut, motha, rush nut, sedge nut, Zulu nut)

The earth nut is widely grown in the tropics, including Africa and parts of Asia (especially India and Thailand). This is also a popular 'snack' product sold either raw or roasted by street stallholders or vendors in India and Thailand.

Eleocharis dulcis Trin. syn. *E. tuberosus* Schult. (Chinese water chestnut, dekang, horses hoofs, potok, wu-yu)

Origin and distribution

This species originates from Eastern Asia and is now widely cultivated in China, India, the Philippines and Indonesia. This aquatic species is cultivated for the production of its swollen shoots which are referred to as the 'chestnuts', botanically these are corms, with a mahogany colour, the corms are otherwise reminiscent of a gladiolus corm. There are local clonal selections but no clear cultivar development.

Crop establishment and production

The Chinese water chestnut is an aquatic plant in the wild, but when cultivated the crop is usually started on the flat, in a plot which can be subsequently flooded from a clear river water source. Ideally the crop is produced in a soil which is high in organic content. The crop is reputed to grow best on alkaline soils although this is not always easy to achieve or stabilize under aquatic conditions.

The selected mother corms are planted direct, or grown on in nursery beds, and subsequently transferred to the final positions when approximately 18 cm high. Either way, the mother-plant material is planted in rows with the corm approximately 10 cm below the prepared soil level and 40 cm apart within the row, with approximately 50 cm between the rows.

The plot is flooded and brought up to field capacity immediately after planting; the plot is then allowed to remain at this level of water availability until the plants are established. The plot is then permanently flooded. The ideal depth of water is approximately 10–15 cm. During the crop's development rhizomes are produced from the corms and enter the muddy substrate, meanwhile aerial leaves develop which emerge above the water level.

Harvesting

The new crop of corms develops in approximately 6 months. Corm maturity is indicated by the initial yellowing followed by dying down of the aerial foliage. At this stage the water supply is stopped and the plot drained. The new crop is vulnerable to mechanical damage and is therefore carefully hand lifted. Following lifting and careful washing the crop is stored in a cool shed or suitable building. Extra care should be taken to avoid damage during the cleaning operation. Sound corms can be stored for up to 10 months.

Further Reading

Tropical grasses

Kelly, A.F. and George, R.A.T. (eds) (1998) *Encyclopaedia of Seed Production of World Crops*. Wiley, Chichester, UK, pp. 271–277.

Wolfe, D.W., Azanza, F. and Juvik, J.A. (1997) Sweet corn. In: Wien, H.C. (ed.) *The Physiology of Vegetable Crops*. CAB International, Wallingford, Oxon, UK, pp. 461–478.

Maize and sweetcorn

Feistritzer, W.P. (ed.) (1982) *Technical Guideline for Maize Seed Technology*. Food and Agriculture Organization (FAO), Rome, Italy.

Appendix 1

Indigenous Species Not Dealt With in the Main Text

Botanical family	Species	Common or local name
Asteraceae (formerly *Compositae*)	*Acmella oleracea*	Toothache plant
	Bidens pilosa	Spanish needles
	Crassocephalum biafrae	Sierra Leone bologi, worowo
	Crassocephalum crepidiodes	Ebolo, fireweed
	Crassocephalum rubens	Yoruban bologi
	Galinsoga parviflora	Gallant soldier, guasca
	Launaea cornuba	Mchunga
	Veronia amygdalina	Bitterleaf
	Veronia hymenolepis	Sweet bitterleaf
Basellaceae	*Basella alba*	Ceylon spinach, Malabar spinach
Capparidaceae	*Gynandropsis gynandra*	Cats whiskers
Cruciferae	*Brassica carinata*	African cabbage
	Brassica juncea	Indian or black mustard
	Brassica nigra	Black mustard
	Erucastrum arabicum	Nechelo, dog mustard, Ethiopian kale
	Nasturtium officinale, syn. *Rorippa nasturtium aquaticum*	Watercress
Cucurbitaceae	*Cucumis africanus*	Wild gherkin, wilde agurkie
	Cucumis anguria	West Indian gherkin
	Cucumis metuliferus	Horned melon
	Cucumeropsis manii	Egusi ito, wet agusi
	Lagenaria vulgaris	Calabash

(Continued)

Continued

Botanical family	Species	Common or local name
	Luffa acutangula	Ridged gourd
	Momordica charantia	Bitter gourd
	Sechium edule	Chayote
Acantaceae	*Asystasia schimperi*	Tropical primrose, Chinese violet
Chenopodiaceae	*Chenopodium album*	Fathen, bathua
Commelinaceae	*Commelina communis*	Asiatic dayflower
Euphorbiaceae	*Cnidoscolus chayamansa*	Tree spinach, chaya
Leguminoseae	*Trigonella foenum-graecum*	Fenugreek, methi
	Crotalaria brevidens	Ethiopian rattlebox
	Vigna spp.	Blackeyed pea, asparagus bean
	Vigna unguiculata var. sinensis	Common cowpea
	Vigna unguiculata var. unguiculata	Catjang
Malvaceae	*Hibiscus manihot*	Sunset hibiscus
	Hibiscus sabdariffa	Roselle
	Hibiscus acetosella	False roselle
	Hibiscus surattensis	Wild sour
Oxalidaceae	*Oxalis latifolia*	Fishtail oxalis
Polygonaceae	*Rumex abyssinicus*	Sorrel, acedera
	Oxygonum sinuatum	Nkenge
Portulacaceae	*Portulaca oleracea*	Purslane, krekot
	Portulaca quadrifida	Chicken weed
	Portulaca lutea	Yellow purslane
	Talinum paniculatum	Fameflower
Solanaceae	*Solanum aethiopium*	African eggplant, scarlet eggplant
	Solanum americanum	Glossy nightshade
	Solanum anguivi	Forest bitterberry, munhomboro
	Solanum macrocarpon	Gboma eggplant, African eggplant
	Solanum scabrum	African nightshade
	Solanum tarderemotum	Black nightshade
	Solanum torvum	Pea eggplant
	Solanum villosum	Red-fruited nightshade
Tiliaceae	*Corchorus olitorius*	Jute, Jew's mallow
	Triumfetta annua	Burweed

The species listed above have been reported to be used as minor vegetables, generally gathered from the wild or allowed to remain in vegetable plots when they occur as 'weeds'. Some of the species listed need careful preparation when used for food in order to remove natural toxins.

In addition to their culinary uses and food value, some of the species have been identified as, or likely to be, potential sources of germplasm in breeding programmes to improve related vegetable species.

Appendix 2

International Research Institutes

The following international research stations are within the Alliance of the Consultative Group on International Agriculture Research (CGIAR):

Africa Rice Center – WARDI, Cotonou, Benin.
Biodiversity International, Rome, Italy.
CIAT – Centro Internacional de Agricultura Tropical, Cali, Colombia.
CIFOR – Center for International Forestry Research, Bogor, Indonesia.
CIMMYT – Centro Internacional de Majoramiento de Maiz y Trigo, Mexico City, Mexico.
CIP – Centro Internacional de la Papa, Lima, Peru.
ICARDA – International Center for Agricultural Research in the Dry Areas, Aleppo, Syrian Arab Republic.
ICRAF – World Agroforestry Centre, Nairobi, Kenya.
ICRISAT – International Crops Research Institute for Semi-Arid Tropics, Patancheru, India with Regional Centres at Bamako, Mali, serving the Sahel area and at Niamey, Republic of Niger, serving Western and Central Africa.
IFPRI – International Food Policy Research Institute, Washington, DC, USA.
IITA – International Institute of Tropical Agriculture, Ibadan, Nigeria.
ILRI – International Livestock Research Institute, Nairobi, Kenya.
IRRI – International Rice Research Institute, Los Baños, The Philippines.
IWMI – International Water Management Institute, Colombo, Sri Lanka.
WorldFish Center, Penang, Malaysia.

Other international centres noted for tropical and subtropical vegetable production activities are:

AVRDC – Asian Vegetable Research and Development Centre, Shan Hua, Taiwan, and AVRDC Regional Centres in Thailand, India and at the ICRISAT Campus, Tanzania at Arusha, Madagascar, Niger, Mali and Cameroon, and also National Partners in Turkmenistan, Tajikistan and Kazakhstan.

Mayaguez Institute of Tropical Agriculture, University of Mayaguez Campus, Puerto Rico.

References

Anonymous (1995) Getting farmers involved in research. *CGIAR News* 2(2), 9–10.

Anonymous (2000) *Cartagena Protocol on Biosafety to the Convention on Biological Diversity*. The Secretariat of The Convention on Biodiversity, World Trade Centre, Montreal, Canada.

Arbizu, C. and Tapia, M. (1994) Andean tubers. In: Hernado Bermejo, J.E. and Leon, J. (eds) *Neglected Crops: 1492 from a Different Perspective*. Plant Production and Protection Series No. 26. Food and Agriculture Organization (FAO), Rome, Italy, pp. 149–163.

Asian Vegetable Research and Development Center (AVRDC) (1998) *Vegetables for Poverty Alleviation and Healthy Diets*. AVRDC, Taiwan, 29 pp.

Awoke, T.C. (1997) The culture of coffee in Ethiopia. *Agroforestry Today* 9(1), 19–21.

Badra, T. (1993) Lagos spinach (*Celosia* sp.) In: Williams, J.T. (ed.) *Underutilized Crops: Pulses and Vegetables*. Chapman and Hall, London, pp. 131–163.

Banga, O. (1963) *Main Types of the Western Carotene Carrot and their Origin*. W.E.J. Tjeenk Willink, Zwolle, Amsterdam, The Netherlands.

Banga, O. (1976) Radish. In: Simmonds, N.W. (ed.) *Evolution of Crop Plants*. Longman, London, pp. 60–62.

Berstein, L. (1970) *Salt Tolerance of Plants*. USDA Agricultural Information Bulletin 283. United States Department of Agriculture (USDA), Washington, DC.

Block, E. (1985) The chemistry of onions and garlic. *Scientific American* 252, 94–99.

Brewster, J.L. (1994) *Onions and other Vegetable Alliums*. CAB International, Wallingford, Oxon, UK.

Brewster, J.L. (1997) Onions and garlic. In: Wien, H.C. (ed.) *The Physiology of Vegetable Crops*. CABI, Wallingford, Oxon, UK, pp. 581–619.

CIP (1979) *Report of the Planning Conference on the Production of Potatoes from True Seed*. International Potato Center (CIP), Lima, Peru, 172 pp.

Cobley, L.C. and Steele, W.M. (1976) *An Introduction to the Botany of Tropical Plants*, 2nd edn. Longman, London.

Cooper, W. and MacLeod, J. (1998) An introduction to genetically modified crops. *Plant Varieties and Seeds* 11, 131–141.

Coursey, D.G. (1967) *Yams*. Tropical Agriculture Series. Longman, Harlow, UK.

Currah, L. and Astley, D. (1996) Describing onions: can the definition of day-length descriptors be improved? In: Currah, L. and Orchard, J. (eds) *Onion Newsletter for the Tropics* 7. Natural Resources Institute, London, pp. 77–81.

Currah, L. and Proctor, F.J. (1990) *Onions in Tropical Regions*. Natural Resources Institute Bulletin No. 35. Natural Resources Institute, Chatham Maritime, Kent, UK.

Dahlquist, R.M., Prather, T.S. and Stapleton, J.J. (2007) Time and temperature requirements for weed seed thermal death. *Weed Science* 55, 619–625.

Diederen, P., Meijl, H.V., Wolters, A. and Bijak, K. (2003) Innovation adoption in agriculture: innovators, early adopters and laggards. *Cahiers d'Economie et Sociologie Rurals* 67, 30–50.

Eponou, T. (1996) Linkages between research and technology users in Africa: the situation and how to improve it. International Service for National Agricultural Research (ISNAR) Briefing Paper Number 31. ISNAR, The Hague.

Fielder, L.A. (1990) Rodents as a food source. In: Davis, L.R. and Marsh, R.E. (eds) *Proceedings of the Fourteenth Vertebrate Pest Conference*. University of California, Davis, USA, pp. 149–155.

Flegmann, A.W. and George, R.A.T. (1975) *Soils and Other Growth Media*. Macmillan, London.

Flemming, J.E. and Galway, N.W. (1995) Quinoa (*Chenopodium quinoa*). In: Williams, J.T. (ed.) *Cereals and Pseudocereals*. Chapman and Hall, New York, pp. 3–84.

Flores, H.E., Walker, T.S., Guimaraes, R.L., Bais, H.P. and Vivanco, J.M. (2003) Andean root and tuber crops: underground rainbows. *HortScience* 38(2), 161–167.

Food and Agriculture Organization (FAO) (1998) *Rural Women and Food Security: Current Situation and Perspectives*. FAO, Rome, Italy.

Food and Agriculture Organization (FAO) (2001a) *Incorporating Nutrition Considerations into Agricultural Research Plans and Programmes*. FAO, Rome, Italy.

Food and Agriculture Organization (FAO) (2001b) *Smallholder Irrigation Technology: Prospects for Sub-Saharan Africa*. FAO, Rome, Italy.

Food and Agriculture Organization (FAO) (2006) *Quality Declared Seed System*. FAO Plant Production and Protection Paper 185. FAO, Rome, Italy.

Food and Agriculture Organization (FAO) (2010) *Quality Declared Seed for Vegetatively Propagated Crops*. FAO Plant Production and Protection Paper. FAO, Rome, Italy.

George, R.A.T. (ed.) (1986) *Technical Guideline on Seed Potato Micropropagation and Multiplication*. Food and Agriculture Organization (FAO) Plant Production and Protection Paper 71. FAO, Rome, Italy.

George, R.A.T. (2009) *Vegetable Seed Production*, 3rd edn. CAB International, Wallingford, Oxon, UK.

Gillman, G.P. and Bell, L.C. (1972) Surface charge characteristics of six weathered soils from tropical North Queensland. *Australian Journal of Soil Research* 14, 351–360.

Gordon, A. (2000) *Improving Smallholder Access to Purchased Inputs in Sub-Saharan Africa*. Natural Resources Institute Policy Series 7. Natural Resources Institute, Chatham Maritime, UK.

Gormley, T.R. (1989) Studies on the role of fruit and vegetables in the diet. *Professional Horticulture* 3, 66–68.

Grubben, G.J.H. (1977) Curcurbits. In: Tindall, H.D. and Williams, J.T. (eds) *Tropical Vegetables and their Genetic Resources*. International Board for Plant Genetic Resources (IBPGR), Rome, Italy, pp. 39–54.

Haldemann, C. (2008) ISTA and biotech/GM crops. In: *Seed Testing International*. International Seed Testing Association (ISTA) News Bulletin No. 136 October. ISTA, Zurich, Switzerland, pp. 3–5.

Hedrick, U.P. (ed.) (1972) *Sturtevant's Edible Plants of the World*. Dover Publications, New York.

Hepper, F.N. (1970) Bambara groundnut (*Voandzei subterranea*). *Field Crop Abstracts* 24, 1–6.

Herklots, G.A.C. (1972) *Vegetables in South-East Asia*. George Allen and Unwin, London.

Hudson, N.W. (1957) Erosian control research. *Rhodesian Agricultural Journal* 54, 297–306.

International Crops Research Institute for the Semi-Arid Tropics (ICRISAT) (1974) *International Crops Research Institute for the Semi-Arid Tropics Annual Report for 1973–1974*. ICRISAT, Patancheru, India.

International Institute for Tropical Agriculture (IITA) (1974) *International Institute for Tropical Agriculture Annual Report for 1972–1973*. IITA, Ibadan, Oyo State, Nigeria.

International Institute for Tropical Agriculture (IITA) (1975) *International Institute for Tropical Agriculture Annual Report for 1974*. IITA, Ibadan, Oyo State, Nigeria.

Katan, J. and De Vay, J.E. (1991) *Soil Solarization*. CRC Press, Florida, USA.

Kelly, A.F. (1988) *Seed Production of Agricultural Crops*. Longman Scientific and Technical, Harlow, UK. Copublished in the USA by John Wiley, New York.

Kelly, A.F. (1994) *Seed Planning and Policy for Agricultural Production*. Belhaven Press, London.

King, L.E., Lawrence, A., Douglas-Hamilton, I. and Vollrath, F. (2009) Beehive fence deters crop-raiding elephants. *African Journal of Ecology* 47(2), 131–137.

Lebot, V. (2009) *Tropical Root and Tuber Crops Cassava, Sweet Potato, Yams and Aroids*. CAB International, Wallingford, Oxon, UK.

MacKay, D.C. and Eaves, C.A. (1962) The influence of irrigation treatments on yields and on fertilizer utilization by sweetcorn and snap beans. *Canadian Journal of Plant Science* 42(2), 219–227.

McDonald, J.A. and Austin, D.F. (1990) Changes and additions in *Ipomoea* section Batatus (Convolvulaceae). *Brittonia* 42, 116–120.

Montiflor, M.O. (2008) Cluster farming as a vegetable marketing strategy: the case of smallholder farmers in Southern and Northern Minanao. *Acta Horticulturae* 794, 229–238.

Montiflor, M.O., Batl, P.J. and Murray-Prior, R. (2009) Social impact of cluster farming for smallholder farmers in southern Philippines. *Acta Horticulturae* 809, 193–200.

National Academy of Sciences (1978) II Roots and tubers. In: *Unexploited Tropical Plants with Promising Economic Value*. National Academy of Sciences, Washington, DC, pp. 29–43.

Nwinyi, S.C.O. and Enwezor, W.O. (1985) Evaluation of fertilizer placement methods for white yam (*Dioscorea rotundata*). *Experimental Agriculture* 21, 73–78.

Onwueme, I.C. (1978) *The Tropical Tuber Crops*. Wiley, Chichester, UK, pp. 122–125.

Oomen, H.A.P.C. and Grubben, G.J.H. (1978) *Tropical Leaf Vegetables in Human Nutrition*. Royal Tropical Institute, Amsterdam and Orphan Publishing Company, Willemstad, Curacao, 140 pp.

Opeña, R.T., Kuo, C.G. and Yoon, J.Y. (1988) *Breeding and Seed Production of Chinese Cabbage in the Tropics and Subtropics*. Technical Bulletin No. 17. Asian Vegetable Research and Development Center, Taiwan.

Organization for African Unity (OAU) (1970) *Symposium on Schistosomiasis*. OAU, Addis Ababa, Ethiopia.

Purseglove, J.W. (1972) *Tropical Crops: Monocotyledons*. Longman, London.

Purseglove, J.W. (1974) *Tropical Crops: Dicotyledons*. Longman, London.

Rao, G.V.R., Rao, V.R., Prasanth, V.P., Khannanal, N.P., Yadav, N.K. and Gowda, C.L.L. (2009) Farmers' perception on plant protection in India and Nepal: a case study. *International Journal of Tropical Insect Science* 29(3), 158–168.

Rawel, K.M. and Kroll, J. (2003) The importance of cassava (*Manihot esculenta* Crantz) as the main staple food in tropical countries. *Deutsche Lebensmittel-Rundschau* 99(3), 102–111.

Revill, P.A., Jackson, G.V.H., Hafner, G.J., Yang, I., Maino, M.K., Dowling, M.L., Devitt, L.C., Dale, J.L. and Harding, R.M. (2005) Incidence of distribution of viruses of taro (*Colocasia esculenta*) in Pacific Island countries. *Australasian Plant Pathology* 34, 327–331.

Robani, H. (1994) Film-coating horticultural seed. *HortTechnology* 4, 104–105.

Robinson, R.W. and Decker-Walters, D.S. (1997) *Cucurbits*. CAB International, Wallingford, Oxon, UK.

Rogers, E.M. (2003) *Diffusion of Innovations*. Free Press, New York.

Salter, P.J. and Goode, J.E. (1967) *Crop Responses to Water at Different Stages of Growth*. Commonwealth Agricultural Bureaux, Farnham Royal, UK.

Sargent, M.J. (1975) *Case Studies in Horticultural Marketing Cooperation*. Bath University Press, Bath, UK.

Sargent, M.J., Wainwright, H. and George, R.A.T. (1987) Horticulture in The Sudan – recent developments, problems and prospects. *Professional Horticulture* 1, 112–119.

Scott, J.M. (1989) Seed coatings and treatments and their effects on plant establishment. *Advances in Agronomy* 42, 43–83.

Shinohara, S. (1984) *Vegetable Seed Production Technology of Japan, Elucidated with Respective Variety Development Histories, Particulars*, Volume 1. Shinohara's Authorized Agricultural Consulting Engineer Office, Tokyo, Japan, pp. 318–426.

Siyal, A.A., van Genuchten, M.Th. and Skaggs, T.H. (2009) Performance of pitcher irrigation system. *Soil Science* 124, 312–320.

Stathers, T., Namanda, S., Mwanga, R.O.M., Khisi, G. and Kapinga, R. (2005) *Manual for Sweet Potato Integrated Production and Pest Management for Farmer Field Schools in Sub-Saharan Africa*. International Potato Center (CIP), Kampala, Uganda.

Stone, O.M. (1982) The elimination of four viruses from *Ullucas tuberosus* by meristem culture and chemotherapy. *Annals of Applied Biology* 101, 79–83.

Thompson, K.F. (1976) Cabbages, kales etc. In: Simmonds, N.W. (ed.) *Evolution of Crop Plants*. Longman, London, pp. 49–52.

Thomson, J. (2007) Genetically modified crops – good or bad for Africa? *Biologist* 54(3), 129–133.

Tindall, H.D. (1983) *Vegetables in the Tropics*. Macmillan Press Ltd, London.

Tonkin, J.H.B. (1979) Pelleting and other pre-sowing treatments. *Research and Technology of Seeds* 4, 84–105.

Tripp, R., Wijeratne, M. and Piyadasa, V.H. (2009) What should we expect from Farm Field School? A Sri Lanka case study. *World Development* 33(10), 1705–1720.

Umaerus, M. (1989) *True Potato Seed*. International Potato Center, Lima, Peru.

United States Department of Agriculture (USDA) (2009) *USDA Foreign Agricultural Service GAIN Report No. E 49002*. USDA, Washington, DC.

van den Ban, A.W. and Hawkins, H.S. (1988) *Agricultural Extension*. Longman Scientific and Technical, Harlow, UK.

Webster, C.C. and Wilson, P.N. (1980) Soil and water conservation. In: Webster, C.C. and Wilson, P.N. *Agriculture in The Tropics*, 2nd edn. Longmans, London, pp. 109–140.

Weinberger, K. and Msuya, J. (2004) *Indigenous Vegetables in Tanzania*. Asian Vegetable Research and Development Center (AVRDC) Technical Bulletin No. 31. AVRDC, Taiwan, 70 pp.

Whitten, M. (2008) Biotechnology and IPM: the relevance to small-scale farmers in Tropical Asia. *Insect Science* 6(2), 97–126.

Wien, H.C. (1997) Peppers. In: Wien, H.C. (ed.) *The Physiology of Vegetable Crops*. CAB International, Wallingford, Oxon, UK, pp. 259–293.

Wilson, D.O. and McDonald, M.B. (1992) Mechanical damage in bean (*Phaseolus vulgaris* L.) seed in mechanized and non-mechanized threshing systems. *Seed Science and Technology* 20(3), 571–582.

Wilson, P.N. and Brigstocke, T.D.A. (1981) The relevance of the animal in vegetable production systems. In: Spedding, C.R.W. (ed.) *Vegetable Productivity*. Macmillan, London, pp. 50–73.

Winter, E.J. (1974) *Water, Soil and the Plant*. Macmillan, London.

Wolfe, D.W., Henderson, D.W., Hsiao, T.C. and Alvino, A. (1988) Interactive water and nitrogen effects on senescence of maize. I. Leaf area duration, nitrogen distribution and yield. *Agronomy Journal* 80, 859–864.

Wolfe, D.W., Azanza, F. and Juvik, J.A. (1997) Sweet corn. In: Wien, H.C. (ed.) *The Physiology of Vegetable Crops*. CAB International, Wallingford, Oxon, UK, pp. 461–478.

World Health Organization (WHO) (1973) *Onchocerciasis Control in the Volta River Basin Area*. WHO, Geneva.

Wrigley, G. (1969) *Tropical Agriculture, the Development of Production*. Faber and Faber, London.

Wurster, D.E. (1959) Air-suspension technique of coating drug particles. *Journal of the American Pharmaceutical Association* XLVIII, 451–454.

General Index

Detailed information on a subject can be found under the names of specific vegetables as well as under the subject itself.

Index of Species